Praise for *The Condor and the Eagle*...

"This book is both intellectually exciting and politically insightful. Holistically it clarifies theoretical physics in the context of its cultural routes. Analytically it introduces the reader to Native American ways of understanding the physical world. It elucidates these two ways of understanding the universe by identifying some fundamental metaphors shared by theoretical physics knowledge and Native American wisdom, and by identifying important differences in their potential influences on the 21st century. The book is authentic in content and stimulating in its implications."

—Dr. Glen S. Aikenhead
Professor Emeritus, Aboriginal Education Research Centre
University of Saskatchewan
Saskatoon, SK, S7H 4V8
CANADA

~~~~~~~~~~~

"Phillip Duran's treatise provides a comprehensive critique and analysis of many of the complex philosophical issues that come into play at the intersection of indigenous ways of knowing and western science. ...he has succeeded in delineating and elaborating on the dialectics of interaction across diverse scientific and cultural arenas in a way that expands our understanding and opens up many new questions and avenues for inquiry. Dr. Duran reaches into the depths of the philosophy of science to offer us new insights that are of benefit to the whole of human existence.

"[His] treatise provides insightful analysis of the deep and complex ontological and epistemological issues that come into play when quite different systems of thought converge. ...a piece of work that clearly extends the boundaries of our understanding of the dynamics at the interface of diverse epistemological traditions, and he effectively makes the case for a more critical and culturally-based approach to the practice of science in the indigenous world."

[Please read the *full comment* under "Reviewers' Comments" at the end of this book.]

—Dr. Ray Barnhardt
Professor of Cross-Cultural Studies, Indigenous Knowledge
Systems
University of Alaska Fairbanks

~~~~~~~~~~

"Indigenous peoples know nature as the inter-relational foundation of their social reality. The natural world is understood as a 'thou,' and not an 'it.' In this way, the natural world becomes a living entity that is contemplated and related to as the source of all life. Using Indigenous worldviews as his conceptual wellspring, Phil Duran has produced a ground breaking thesis for a 21st century Philosophy of Science that enables the possibility for the integration of an ecologically based and ethically informed relationship to the natural world and the practice and application of science. 'Can scientists embrace the sacred web of life without compromising the integrity of their discipline?' Phil Duran's answer is Yes! And, as an affirmative response to this essential question, this work is destined to become a corner stone for the 're-indigenization' of science for Indigenous peoples. Simultaneously, it has the potential for the kind of 'revitalization' of Western Science practice that is so critical in our socially and environmentally challenged world. I recommend *The Condor and the Eagle to* Indigenous scholars and Western scientists alike as a guiding epistemology for a new and truly integrated approach to science that is so desperately needed in our climate challenged world."

—**Gregory A. Cajete**, Ph. D
Santa Clara Pueblo
Director of Native American Studies
The University of New Mexico

~~~~~~~~~~

"Duran's book offers a new paradigm in which expressions, knowledge, and social justice realities are free of religion. All readers need to know that 'Western values have contradicted the very holism and authoritative principles inherent in nature and revealed by modern physics.' This is a reaffirmation that Indigenous people have always held; that the proper way to live is linked to the understanding of and respect for nature.

"Among Indigenous North Americans we need more discussion about 'our living in a world of illusions, unaware of the unseen world of reality that supports the whole universe.' We need to discuss the lack of connection between knowledge and morality and the strong connection between knowledge and power and the political influence of wealth...

"May all readers be blessed with Duran's awareness 'of a universe that is endowed with its own consciousness, that is intelligent, authoritative, and harmonious – the manifestations of a spiritual reality from which the Creator made all things and to whom we are responsible.' What we have here in this book is a spiritual, scientific prayer that is years ahead of its time and will surely lead the way to a moral, spiritual and intellectual consciousness."

[Please read the *full comment* under "Reviewers' Comments" at the end of this book.]

— **Dr. Vivian Delgado** (Yaqui)
Professor of Indigenous (Indian) Studies
Bemidji State University

~~~~~~~~~~~

"Einstein said that the business of science is 'reality'. But, whose reality? To a large extent, what is commonly referred to as 'science' has largely been monopolized by Western cultures. The view of science, from a Western perspective, is one of measurement based on mathematics. But there are other approaches to the human search for reality...

"The seeing of a very different version of 'reality' is one reason why North American Indians are forever explaining themselves to Westerners. ...Native Americans become wonderful and great storytellers because they introduce to Westerners ideas that have never crossed their minds.

"In my many nomadic wonderings, physically over 'turtle island' and academically, I am always in search of Native Americans who have a special interest in science, especially in quantum physics. ...Phil Duran is a great storyteller who, as a physicist, was very brave to cross the boundary between two scientific worlds. He is a true warrior, who is brave enough to cross the boundary into the Western physics world to explicate the Native American scientific way. I am twice blessed: I found another Native American scientist steeped in quantum physics and *THE*

CONDOR AND THE EAGLE is a wonderful story about boundary crossings between Western and Native American sciences. It is a 'must' read for scientists, both Western and Native American."

[Please read the *full comment* under "Reviewers' Comments" at the end of this book.]

—Dr. Leroy Little Bear
Blood Tribe of the Blackfoot Confederacy
Former Director, American Indian Program at Harvard University
and
Professor Emeritus of Native Studies at the University of Lethbridge

~~~~~~~~~~~

"*The Condor and the Eagle: Uniting Heart and Mind in Search of a New Science Worldview* is a book that is perfect for our time. I also believe that 'Phillip will become a great scientist,' because of this new ground-breaking world view. Now for some time we have been experiencing climate changes on our Mother Earth. The Condor (heart) and the Eagle (mind) are part of all of the changes we will be experiencing as we the people move into a new world of abundance and enlightenment."

—Joseph Rael
Picuris Pueblo/Southern Ute
MA Political Science, 1976, Wisconsin Madison

The Condor and the Eagle

Uniting Heart and Mind in Search of a New Science Worldview

PHILLIP H. DURAN

© 2013 Phillip H. Duran

The Condor and the Eagle: Uniting Heart and Mind in Search of a New Science Worldview

ISBN-13: 978-0615815510 (Eaglehouse Publications)

Phillip H. Duran
phil-duran@cougarmail.org

Front cover art by Dennis Arneson. Prophetic condor and eagle flying together wing-to-wing over the Americas; atom; semi-transparent galaxy and colors of the four directions in background—all in motion.

Eaglehouse Publications
Rio Rancho, New Mexico

For Worldwide Distribution

TABLE OF CONTENTS

FOREWORD

Richard Simonelli

Native people acknowledge the eagle as an inspiration and keeper of spirit energy on Turtle Island, the Native name for the North American continent. Likewise, in the southern lands of the Western Hemisphere the condor holds a similar role as a keeper of spirit energy in South America. There are Incan teachings and prophecies from South America that speak of the great division between the indigenous peoples of North and South America that began over 500 years ago with arrival of the European colonizers. These teachings say that the encounter 500 years ago not only separated the continents because of Euro-centric control, but also brought about an imbalance in masculine and feminine energies everywhere through patriarchy. Euro-centric ways imposed an increasing disconnect between the hearts and the minds of all peoples. They said that this growing division and this split would result in great suffering on planet earth, which we do see happening now.

The teachings also noted that during these times of separation the condor and the eagle would no longer fly together as they once did, further expressing disunity in a long 500-year wintertime. But there is also an ancient Incan prophecy saying there would come a time of reunion and unity in the distant future when the condor and the eagle would fly together once again – a new springtime. At that time, the divided masculine and feminine energies on earth would begin to reunite and the split between the heart and the mind in all peoples would begin to heal. This is one of many different interpretations possible for the eagle-condor prophecy. We are now in that coming-together time foretold so many years ago.

The Condor and the Eagle: Uniting Heart and Mind in Search of a New Science Worldview is a book expressing the unifying condor and eagle prophecy as it is beginning to play out in Western science, which has also suffered a divided or dualistic thought process over the centuries. Through quantum physics and the holistic quantum worldview that began to emerge starting in 1900, the Western scientific process may itself begin to find wholeness in this new coming together time.

The Condor and the Eagle is a book noting that the worldview of traditional tribal peoples is one of harmony, balance, wholeness and interconnectedness similar to that which may be attributed to the findings of quantum physics. The book teaches that humankind needs to embrace, learn from and begin to live through the holistic worldview of the Native peoples still among us if we are to deal creatively with our problem-ridden future. Increased human conflict, war, hate, corporate greed, nationalism, overpopulation, a toxic natural environment, environmental illnesses, global warming, escalating weaponry – these are some of the unfortunate legacies to our grandchildren. This pattern of conflict and distressed relationships is first and foremost an outcome of a divisive worldview. *The Condor and the Eagle* is a book that says we need to put behind us our divided, dualistic, separated worldview, which is also at the root of a classical physics mindset (specifically, reductionism and mechanism), if we are to survive as human beings with sustainable societies in a survivable future.

This is a book about worldviews and ways *of knowing* which underlie science and technology even if we might not be specifically aware of them. It's a book about Native peoples, quantum physics, the new science and a sustainable future for our grandchildren. Author Phil Duran takes us through the

Native American experience in North America to show how a divisive worldview and foundational philosophy separates, rejects and hurts rather than includes, harmonizes and heals. He shows us that the roots of Newtonian physics and the philosophy of René Descartes, key to Western life, are expressions of a separative worldview. He takes us through the development of classical Western physics until it discovers the quantum in 1900 and the entire game begins to change.

Author Phil Duran is both a Native American and a physicist with a career in physics, physics education, and information technology at the mainstream and tribal college levels. He's got one foot in each world but his passion is one of unification so that two worlds may be revealed as one. In these chapters he speaks of the historical Doctrine of Discovery that was used by 15th Century Europeans to justify the conquering and theft of land from Native people by proclaiming European superiority and divine right to displace those of a different orientation. Even though the Doctrine of Discovery is an artifact of historical times its divisive mindset is still alive today in American foreign policy and in the attitude of a conquering globalization. It's an expression of a hubristic worldview that needs to change if we are to live a sustainable future.

These are difficult times for human beings and the natural environment alike. They are also dispirited times because so many people seem to have lost their sense of fundamental meaning, purpose, wonder and spirituality, which are like compass directions to take us forward. But for many people today, the findings, interpretation and vision of quantum physics is one bright light of inspiration for themselves and their families, friends and colleagues. But how many people really know what quantum discoveries and their underlying physics really mean? Utilizing his experience as a teacher, author Phil actually discusses, presents and teaches the key points and ideas of Einstein's relativity and the quantum universe in this book. He tells the quantum story clearly and accessibly. What are some of these stories in the book?

- Discovery of the quantum in 1900
- Wave-particle duality: the photoelectric effect
- Einstein's special and general relativity
- Pauli's exclusion principle
- De Broglie's waves
- Heisenberg's uncertainty principle
- The separate "observer" becomes the "participator": *the observer is the observed*
- The quantum mechanics wave equation
- The many worlds interpretation of quantum mechanics: parallel universes
- Schrödinger's cat
- Neils Bohr's Copenhagen interpretation
- The Einstein, Poldosky, Rosen paradox of 1935: classical locality
- Bell's theorem of 1964 and Alain Aspect's experimental verification: reality is non-local!
- "M" or string theory
- David Bohm's "Wholeness and the Implicate Order": the undivided universe
- Paul LaViolette's primordial "substance"
- Consciousness and quantum reality

And so much more. The reader can go deeply into these topics and use them as jumping off points for further research, or simply summarize them to get the overall flavor. He or she may take an eagle's or condor's "top view" of the stories of quantum physics and the new science as needed. The reader may also reflect on the important issue of context: when is use of the older classical viewpoint appropriate, and when (and how) should quantum thinking guide our lives?

For myself, some years ago it was my discovery of the implications of Bell's Theorem and its practical proof through Alain Aspect's 1981 experiment that really blew my mind. Bell's theorem is a mathematical statement that can be summarized as, *No physical theory of local hidden variables can ever reproduce all of the predictions of quantum mechanics.* It was validated by experiment in 1981 when French physicist Alain Aspect showed that two photons were intimately related through quantum entanglement over a physical distance of some thirty feet by experiments in his lab. In other words, they were "in communication," or "they knew each other" without the need for a signal to pass between them. Perhaps they were "one" in quantum wholeness. Phil's book reveals that in later experiments this quantum entanglement was demonstrated over distances as much as 27 km.

Bell's theorem can be interpreted as establishing that *reality is non-local.* In other words, deepest reality does not consist of separate things acting at a distance on other separate things. Non-local can also mean non-dual: not two separate things. Reality is non-dual, or perhaps *one.* The wholeness of the indigenous worldview is at last discovered by quantum physics. It blew my mind because it was akin to my own experience in meditation, as well as experience I had in sweat lodge and through ceremonies with the Sacred 100 Eagle Feather Hoop of the Wellbriety Movement. For many years the Wellbriety Hoop had a condor feather hanging at the center of its Hoop of 100

Eagle Feathers to ceremonially join the condor and the eagle in fulfillment of prophecy.

As a supporter and active participant of the Wellbriety Movement for healing of Native Americans (inspired by Don Coyhis of White Bison, Inc.), I am privileged to hear our elders' statements expressing the holism of the indigenous worldviews, which is so necessary now. For example:

- The honor of one is the honor of all, and the pain of one is the pain of all.
- The longest road we will ever walk is the 18" from the head to the heart.
- We look towards and become like that which we think about. If it is true that we become like that which we are thinking about, isn't it time we began to think about what we are thinking about?

The Condor and the Eagle is a book for science professionals of all backgrounds. It is a book that will find its way into Native studies curricula, into anthropology, philosophy, history of science, environmental studies and education departments. It is a book both Native people and scientists will recognize. It is a book for general readers who would like to connect indigenous and quantum reality. And It is a book to help us understand how we got to where we now stand in this profound but also very perilous time on the good mother earth.

Author Phil was moved to include in the book this statement from the Thirteen Indigenous Grandmothers, who have been criss-crossing the earth in their travels with a message of sanity and healing. I'll end here with these indigenous voices:

"We, the International Council of Thirteen Indigenous Grandmothers, believe that ancestral ways of prayer, peacemaking, and healing are vitally needed today. We come together to nurture, educate, and train our children; uphold the practice of our ceremonies, affirm the right to use our plant medicines free of legal restriction, protect the lands where our peoples live and upon which our cultures depend, safeguard the collective heritage of traditional medicines, and defend the Earth Herself. We believe that the teachings of our ancestors will light our way through an uncertain future."

Richard Simonelli, MSEE
Nederland, CO
May 18, 2013

Formerly contributing editor, *Winds of Change* magazine,
(American Indian Science and Engineering Society)
Media Specialist, White Bison, Inc.
Editor, Coyhis Publishing, Inc.

The Condor and the Eagle

PREFACE

Phillip H. Duran

One memorable day in the spring of 1955—the last semester before my graduation from high school—our English teacher asked us to tell how we envisaged the professional future of other students in the classroom. One of them said, "I think Phillip will become a great scientist."

Coincidently, Albert Einstein passed to the spirit world in that same semester (April 1955) at the age of 76—my age when this book was written. Of course, the student's prediction didn't come true but her comment was an encouragement nonetheless. In fact, I chose physics as my major when I entered The University of Texas at El Paso[1] that year. After earning a B.Sc. degree in physics, I was accepted for graduate work at The University of Iowa[2] in the department of physics to study astronomy. The chair of the department, James Van Allen, was a gracious man who spoke kindly and cared about me as a student. We talked about the future of the doctoral program in astronomy, whose main professor had moved to Australia. When he learned that I was financially stressed, he gave me a job working in his Explorer satellite program. But after two months I discovered that I was not sufficiently strong in mathematics to continue in the graduate physics program, so I dropped out and came home.

I still get nostalgic when I think about that Iowa experience; although my time there was brief, the people and events represented a meaningful experience that was taking me to

---

[1] Previously Texas Western College

[2] Previously the State University of Iowa (SUI) in Iowa City

unknown realms in association with a space scientist of renown; it also represents a truncated path that I never completed.

I soon returned to the UTEP campus to work in a physics research lab. In 1961, I began teaching elementary courses in physics and mathematics; later I taught advanced courses while working part-time on a M.Sc. degree in physics, which I earned in 1970. For my master's thesis, my advisor, Thomas G. Barnes, asked me to build a mathematical model of sound propagation in the upper atmosphere in order to compute wind gradients analytically instead of the current method that required White Sands Missile Range to detonate grenades from rockets fired into the upper atmosphere.

The problem was important to the U.S. Army because such a model would utilize explosives on the ground and eliminate the need for rockets. The model I developed made it theoretically possible to calculate the required sound intensities, contingent on obtaining an *analytical* solution to the equation; *i.e.,* without using approximation methods. This was difficult because the equation was an integral with complex variables— Bessel functions of imaginary order and imaginary argument. Moreover, since the denominator generated infinities, or singularities, it had to be solved as a separate problem before the integral could be solved.

I succeeded in solving the equation using the function in the denominator as the problem for my master's thesis; this produced the zeros needed to solve the complex integral and, since my work was funded by a grant from the U.S. Department of the Army, I also submitted a technical report.[1] Four months later, I solved the complex integral equation and wrote a second technical report,[2] for which I received a citation from the Department of the Army stating that my work was a significant achievement of specific relevance. At my advisor's request, R. W. Stevens, a member of the Royal Society of London and a recipient of the Rayleigh medal, reviewed my report and assessed it as Ph.D. level work.

In 1971, my wife and I and our three children moved to Pullman, Washington so that I could begin doctoral studies at Washington State University in the Department of Computer Science. After seven strenuous years, having passed the qualifying exams and the preliminary oral exam in computer science and theoretical physics, I was stunned to learn from my Ph.D. advisor that he had submitted a paper to publish his results on the same dissertation problem that he had assigned me. Our results were identical but mine were useless. He said he had done it to test my ability to perform original research. He not only doubted my ability (which I had already demonstrated through research at UTEP) but also committed a fraudulent act. Efforts to continue in the doctoral program were futile. I earned another M.Sc. degree and, since I could not find employment anywhere without a Ph.D., I spent the next twenty years at WSU working as a systems analyst.

After my father crossed over in 1991, I learned that our family on both the paternal and maternal sides are related by blood to the Tigua Indians (Ysleta del Sur Pueblo), a tribe located in El Paso, Texas—a fact that neither of our parents had made known to their three sons. Almost immediately, my search for identity began: the meticulous examination of census and church records, the endless study of American Indian history, and learning about the ongoing issues in Indian country. The experience was long and difficult but also valuable and necessary. My wife and I began to spend much of our time in the tribal communities among the Plateau tribes of the Pacific Northwest. We also became campus and community activists; I wrote a number of newspaper editorial columns and hosted documentary film screenings on the WSU campus.

In 1998, I became director of a new college-level institute in South Dakota, designed specifically for American Indian students and, although it closed down two years later due to insufficient funding, the experience of working directly with Indian students and the direct experience of relating to members of the Oglala Lakota Nation were essential to me as the continuation of an indigenization process.

In 2000, I became director of an environmental studies degree program in the Division of Science and Mathematics at Northwest Indian College, where I later served as its dean. I also taught courses in conceptual physics and American Indian history. I refreshed, re-learned, and studied quantum mechanics and relativity theory—the post-Newtonian concepts that comprise modern physics[3]—from an *Indigenous*[4] perspective. For me, the domain of physics was no longer limited to the material universe; I now contemplated a live physical/spiritual Universe and perceived physics through Indian eyes, linking relevant concepts in science to Indigenous cosmological concepts that involved questions about the validity of experience, consciousness, the nature of reality, time and space, matter-energy transformations, the relationship between energy and spirit, cosmology, and more.

---

[3] M-theory (formerly string theory or superstring theory) is also post-Newtonian but not discussed in this book.

[4] The terms *Indigenous* and *Native* are used interchangeably throughout this book; this is a common practice among Native authors. A *people* is a plurality of persons considered as a whole, as in a single tribal nation. When referring to more than one people or nation, the term *peoples* is used. Chapter One, Article One of the Charter of the United Nations states that [all] peoples have the right to self-determination. This right extends to Indigenous peoples—the world's original inhabitants and their descendants who have suffered colonization. The following terms refer to Indigenous peoples from the Americas: Indigenous, Alaska Native, Aboriginal, Native Hawaiian, First Nations (Canada), *indio*, Native, Native American (U.S.), American Indian (U.S.), Indian (U.S.). In 2007, the United Nations General Assembly adopted the Declaration on the Rights of Indigenous Peoples.

The Condor and the Eagle

After spending thirty-four years in the Pacific Northwest (except for two years in South Dakota), my wife and I moved to New Mexico—the place of my origin among the Indian Pueblos—to continue the journey of discovery and identity. My wife and I spend most of our time with Indian families and we frequently visit Indian communities. Much as we would like to have our extended family close by, as it was during my childhood, we have to accept that times have changed but we're fortunate that our children and grandchildren remain in frequent contact with us.

Despite seventeen years of graduate coursework and additional time spent in self-study, I have forgotten most of my mathematical skills due to my twenty-year absence from teaching and research except for a physics course that I taught in 2001. Although I have no active research area and no longer teach or practice mathematical physics, it never leaves the soul of a physicist. And despite extensive "book learning," I am an Indian by blood and in my soul, though the effects of colonization do not altogether disappear.

This book represents a part of my contribution to the world in lieu of what I might have offered as the "great scientist" that I never became. The Indian voice and the voice of science both speak on these pages; in both cases I speak from my own Indian perspective and identity. As an Indian, I sense complete freedom to invoke Creator and his Universe. But as a physicist, I feel somewhat constrained. As a Native physicist—a physicist with an Indigenous perspective—I perceive and experience, with feeling, both the spiritual and the physical.

Physics concepts may be boring and dull to some readers because these truths may not invoke emotion; they challenge our intellectual powers and our capacity to interpret abstract concepts. I believe the key is to contemplate the Universe with both mind and heart together.

A Word about Terminology

Academic terms like *physics* and *science* have the usual meaning in this book except when the context indicates differently. If we say "physics is the study or investigation of the properties of matter," it would fit the usual meaning. Likewise, *science* and its derivatives, such as *scientific* and *scientific method*,[5] are references to Western science or its empirical methods. But the phrase "the people's own science" clearly means the people's own system of knowledge, not the usual meaning.

Regarding the word *nature*: In ancient Greece, the investigation of nature was metaphysical; Greek physics differed from our modern meaning because the earliest Greeks believed that nature has life and spirituality. *Nature* also has a different meaning to Indigenous peoples compared to how it is understood in Western thought. Perhaps the single, most helpful advice to the reader is to think of Indigenous words in terms of verbs rather than nouns; this will ascribe a dynamic, flowing, and creative characteristic to the Indigenous meaning.

Interpretations of physics concepts and Indigenous perspectives are my own. I do not speak for any tribe and there is no attempt or intent to prove or interpret Indigenous beliefs or spirituality using physics. The emphasis on modern physics was deemed necessary to provide a contrast from the classical and illusory everyday world and to give readers an adequate overview of a topic with which they may not be familiar. I also want to give readers who are not very familiar with physics much more than a vague understanding of quantum theory and

---

[5] Some textbooks define "scientific method" as a specific sequence of steps and was first propounded in the 16th century to ensure that rational thinking and experimentation are used in the validation of new knowledge.

The Condor and the Eagle

relativity theory. But I do not mean to imply that modern physics is itself a holistic path.

The Central Message of This Book

The following statement expresses the main theme and premise of this book:

> Modern physics describes our Universe as a world of interconnected networks in which all things are interrelated, interdependent, and interconnected. It reveals facts and mysteries about the Universe that are spiritually significant to Indigenous peoples, who have known and practiced them for millennia, while contrasting with Western culture's worldview and practice of science. It is also significant that some of the early Greek masters of wisdom acknowledged and taught that nature is alive and endowed with spirituality—a belief also held by Indigenous peoples. From these evidences it can be inferred that nature is pointing to a holistic path which the West abandoned long ago. Modern physics is not that path; it is the tool which nature has used in pointing the West to a path already followed by Indigenous peoples. In response, the West needs to recognize and act consistently with the obvious link that exists between the truths which nature is conveying and the proper way to live. In recent years eminent physicists have written popular books for a general audience, explaining the concepts and principles of modern physics, yet none of them have so far linked together these important truths as I do in this book.

Acknowledgments

I am indebted, first of all, to Norma, my wife and lifetime companion, for her constant support and assistance; to our precious sons and daughter and their families; to the Native scholars who have taught me many things both in person and through their books and have invited me to participate in planning meetings across the country, including Alaska, and who collaborated with me in various other ways; to our friends and elders in the Umatilla, Choctaw, Nez Perce, Lakota, Lummi, and Pueblo tribal communities; to Glen Aikenhead, Richard Simonelli. Joseph Rael, Vivian Delgado, and Ray Pierotti for their helpful advice during the preparation of the manuscript; to numerous friends from many tribes who shared their histories and were generous with their time; to those everywhere whom I have not met but in the cause of justice have dedicated their lives and persevered under difficult circumstances working to heal of our Mother Earth; to the world's elders who have preserved and shared their wisdom and knowledge for all generations.

Opening Prayer

An honor song and an offering have gone up to the Great Spirit, who makes our hearts glad as we return some measure of what we have been given. In this spirit this book was written; in the same spirit it is offered to the world.

Accept our honor song and strengthen our faith in your power to heal. Give the leaders of a divided nation and a sick world the moral courage to expose wrong and do what is right. Give them listening ears to hear the Indigenous voice. Make our hearts strong and bring pure light to our minds to serve the people wisely.

Phillip H. Duran
Rio Rancho, New Mexico
April 2013

The Condor and the Eagle

REFERENCES

[1] Duran, Phillip H. *Evaluation of Bessel Functions with Imaginary Order for Applications to Certain Boundary-Value Problems*. Final technical report for 1 Jul 69-30 Jun 70, accession number AD0728833, The University of Texas at El Paso, June 1970.

[2] Duran, Phillip H. *Sound Intensity in Atmospheric Shadow Zones Assuming a Constant Velocity Gradient*. Final technical report for 1 Jul-30 Sep 70, accession number AD0728836, The University of Texas at El Paso, December 1970.

INTRODUCTION

For more than a decade I have reflected on the relationship between modern physical cosmology and the knowledge and beliefs of Indigenous peoples, who have always embraced a land ethic because the Earth sustains all life. The connection to the land is what renders this relationship crucially important at this time in human history because of its potential to inspire and unite the world's peoples and governments in a common cause that focuses on the welfare of future generations as well as our own.

One of the premises of this book is that there is a link between a correct understanding of nature and the proper way to live. This long-held belief of Indigenous peoples is reaffirmed by contemporary American Indian writers and was also embraced by some of the early Greek spiritual masters in the 5th century BC, particularly Empedocles.

The title of this book was inspired by the fulfillment of the Prophecy of the Eagle and the Condor, which foretold that the Indigenous peoples of the northern and southern parts of the Western Hemisphere would be separated for 500 years, beginning with the European invasion of 1492, after which they would be reunited, when the two peoples will have the opportunity to share their knowledge with each other. That time has arrived; the 500-year period, which some have called "the bitter spiritual wintertime," is over.

By inference, as I explain later, the relationship mentioned in the foregoing paragraphs is an application of this prophecy; it is the union of mind and heart—the mind of the eagle and the heart of the condor—as embodied by Western science and

Indigenous knowledge, respectively, to create a new, holistic paradigm in science. But because of the hegemony and arrogance of Western culture, it is the West which needs to become receptive to the knowledge and wisdom of Indigenous peoples. The eagle needs the heart of the condor; Western culture direly needs to change from a legalistic and rationalistic focus to a broad-minded attitude that is compassionate and holistic in its approach to problem-solving in all areas of life.

Within these pages you will also discover a new way of looking at physics. You will see how the quantum and relativity theories reveal the beauty, intelligence, and spirit with which the Creator imbued nature; how intertwined, complex, and alive our Universe is; how true it is that "all things are connected" and that "we are all related"—qualities and principles of sustainability that the world's Indigenous peoples have known and practiced for thousands of years through a loving relationship with nature.

The Greek word for nature is *physis*, from which the word "physics" is derived; it had a much broader meaning in ancient Greece compared to our modern understanding of nature. Some of the earliest Greeks who preceded Socrates believed that nature is endowed with life and spirituality, that all things are vibrantly and knowingly alive—a view that is also widely held by Indigenous peoples. The discovery[1] in 1900 that the transfer of energy occurs in bundles or packets called *quanta*, followed by additional discoveries in the decade that followed, revolutionized physics and raised new philosophical questions about the nature of reality, matter, energy, time, and space. These discoveries also essentially destroyed the reductionist

---

[1] The term "discovery" occurs frequently throughout this book to convey the notion that something new has appeared which was not noticed before—the sudden or gradual awareness of a new idea, invention, or fact. In this sense, the *quantum* and numerous laws, principles, and theories are also discoveries.

view of the Universe—that it can be broken down into its smallest components (consisting of inert, indivisible bits of matter), analyzed independently, and put back together like a machine. This leaves as the only possible alternative the conclusion that the fundamental physical constituent of the Universe, since it is not matter, must be energy (or the vibrating "strings" of M-theory). We also know, of course, from the famous Einstein formula, $E = mc^2$, that matter and energy are equivalent. But the quantum's strange behavior has also spawned divergent philosophical interpretations, moving science into an area that borders on mysticism, which is a taboo in physics. These considerations easily accommodate the Indigenous concept that all things ultimately stem from spirit and that the Universe is alive.

Discoveries in modern physics have revealed even more surprises. One is that the Universe is a highly interconnected and undivided whole, like a tapestry or a web of relationships and the other is probably the most amazing yet well-established natural phenomenon: Two particles[2] which may be far apart, even by millions or billions of miles, can be mysteriously linked together such that whatever happens to one of them immediately causes a change in the other one. The twin particles, said to be "entangled," are forever connected or correlated somehow as if they belong to a single system or unit, yet there is a large distance between them. This phenomenon, called "non-locality," has been confirmed repeatedly in experiments and, although as yet unexplained, it implies that the Universe is deeply connected in some mysterious way. As it is often said in Indian country, "all things are connected."

If we have learned anything from nature through modern physics, it should be a universally acceptable perspective to

---

[2] In the context of physics a material *particle* is a small object whose dimensions are considered insignificant.

address the crises we face nationally and world-wide—a perspective that hasn't been noticed because something has blinded the industrialized world for more than two thousand years. The fundamental problem is that *Western science separates humans from nature.* Instead of seeking a direct and proper relationship with nature, alienation became a chronic human condition long ago.

My own response uniquely and fundamentally addresses this problem by appealing to the human conscience in search of a worldview that everyone can embrace. By providing an overview of quantum theory and Einstein's relativity theories and how they relate to Indigenous worldviews, proven principles of sustainability that have been practiced for thousands of years will become visible and hopefully acknowledged as viable concepts for implementation by mainstream science.

Environmental efforts which have been underway for decades in the United States can be viewed as the first step in an intentional and public process which views physics from a new perspective to help us re-discover our place in nature and to recognize our individual and collective responsibility in the sacred web of life. We must, however, be careful not to turn physics into a cultic ritual based on erroneous ideas about quantum theory. Physics must retain its genuine character while interpreting phenomena in ways that are sound and feasible.

I also hope this book contributes meaningfully to the re-indigenization[3] efforts of Native people. The world our ancestors knew was transformed by European ways into one that is vastly different from the way it was meant to be according to the Creator's original instructions. One of the

---

[3] Re-indigenization refers to the efforts of Native people to revitalize their cultures (Native languages, foods, ceremonies, land), physically and spiritually, in order to rebuild what has diminished since first contact with Europeans.

pervasive characteristics of Western thought, which is evident in all Western societies, is the absence of the sacred from public life. Except as practiced by individuals, the sacred is missing from science, public education, and most other activities that are not church-related.

Living under the laws of the imposed government of a foreign culture, however, does not imply that we must forfeit our own culture and worldviews, especially when the colonizers, now more than ever, need the same Indigenous wisdom that they have spurned. From time to time, members of a tribe need to come together, bonded by their common spirituality as a people, to focus on ceremony at a place away from all outside influence. My friends, the Chickamauga Cherokees of the Mid America Indian Fellowships, do this regularly, calling it a "refuge" from the dominant culture.

Quantum mechanics obliges the physicist to interact subjectively with nature before a particle's property can be measured, implying that the physicist (the observer at the macro scale) must somehow communicate with the particle (the observed at the micro scale). This question of observer versus observed was so important to physicist David Bohm that, when he heard about Jiddu Krishnamurti, a renowned spiritual teacher who had written about it from a philosophical perspective, he made arrangements to meet him personally.[4]

The world has experienced numerous movements that have come and gone. Governments, organizations, grassroots movements, and individual families are all participating in the "Age of Environmentalism," as our time in history has come to be called, judging from the titles of numerous books about the environment, waste reduction, clean energy, a sustainable future, and similar issues. Global climate change should be a top

---

[4] The Krishnamurti-Bohm collaboration is further discussed in chapter 6.

priority but, unfortunately, in the United States, it is a political issue influenced by partisan differences.

Like other movements in the history of the world—the Age of Enlightenment, the Renaissance, the Romantic Era, the Industrial Revolution, the Atomic Age, and other significant periods in history—modern environmentalism is a *reaction* by a section of the general population to the widespread conditions they oppose or to those they deeply desire either to bring back or create anew. New poetry, music, novels and other books, theatrical dramas, and other creativity reflect the activities within each movement and add new terminology to the language and the symbolism associated with the era.

I know little about the causes and patterns of social movements but I speculate that in most cases a movement begins when conditions are right and all that is needed is a "spark" to light it or a "push" to set it in motion. Pollution and the use of pesticides inspired Rachel Carson to write *Silent Spring*, a book published in 1962, which helped launch the American environmental movement. Eventually, new branches of science were formed and new practices and products in technology were created.

For example, "green" is now the environmental symbol that defines the right principle, product, idea, or science; we have green and clean technology, sustainable development, environmental science, photovoltaic cell technology, wind and solar power, and recycling. *Green chemistry* consists of chemicals and chemical processes designed to reduce or eliminate negative environmental impacts, resulting in reduced waste products, non-toxic components, pollution, and improved efficiency; it is not just a theory but applies to real-world environmental situations. Principles for manufacturing green buildings include energy efficient architecture, efficient (or reusable) materials and resources, water conservation, indoor environmental quality, and site and community impact.

We must not think, however, that unless governments and organizations are doing something about the environment, nothing is being done. I made this mistake until I came upon Paul Hawken's book, *Blessed Unrest*, where he describes the largest movement on Earth, comprised of individuals and groups working on environmental causes; it is not organized and has no name, leader, or contact information, yet it reflects brilliant ideas, novel strategies, and people behind them who cannot be identified.[1]

Being "close to nature" involves more than visiting the national parks and spending more time outdoors. This will not restore a broken relationship. Recycling, water conservation, replacing light sources with fluorescent bulbs, organic farming, green buildings, and making global warming a national priority (which is *always* subject to decline due to more immediate crises) are crucial but, as already mentioned, they can only be the first step in a permanent strategy. And to better understand the meaning of Indigenous knowledge, read Wade Davis's account (*The Wayfinders*) of how the Polynesians can make long voyages without modern instrumentation, relying only on the fundamental elements of the Polynesian world: wind, waves, clouds, stars, sun, moon, birds, fish, and water.

From the beginning of the environmental movement until recently, the term "environment" has connoted the wrong idea: that nature surrounds us and that we are separate from it. Environmental advocates eventually realized that humans are a part of nature. Nonetheless, eons before "environment" came into usage, this so-called "wrong idea" is what separated humans from nature, which helps to explain why our world may now be facing the worst crisis in history. As I describe in chapters 1 through 3, the crises are part of the legacy of Western civilization; humans have exploited the natural world as if they were *not* a part of it, assuming that "progress" — the advance of civilization focusing exclusively on human interests — at Earth's expense would not eventually affect future generations. It took modern physics (*i.e.*, nature) to teach us that, indeed, as Native peoples have been saying all along, "all things are related"; the life-sustaining principle of the interdependence of all things has been violated for over two thousand years and, owing to the complexity of the issues, adequate solutions seem far from reach. The *environmental impact* is evident. The richness of the soil, the purity of the waters, the clean air that our ancestors enjoyed — everything changed as the inevitable outcome of a relationship that is hostile toward the earth. The *political crisis* is also evident. In the United States, the national debt and national deficit are in terms of trillions of dollars. Corporate greed and arrogance are still winning, and we are involved in several wars. It is a culture of over-consumption, hostility, and greed, driven by the deeply engrained values of Western culture: take more than you need; the earth belongs to us; tame the wilderness; consume the "resources," and make a profit.

The Western World has long prided itself about its science, even to the point of arrogance, reluctant to acknowledge the casualties that stemmed from the abuse of its powerful science and technologies. Because the West is largely oblivious to the knowledge of Native peoples, not to mention the numerous ongoing issues that the latter face, it has not known how to benefit from the multiple centuries, even millennia, that Native peoples have lived in respectful relationship with Mother Earth.

This past decade also witnessed a rise in the significance of globalization in all its aspects: communication, economy, war, and natural disasters. Because of the dependencies between U.S. and European banks, the economic meltdown experienced by the U.S. in 2008 was felt internationally.

On the social front as well, we seem to be losing ground. Poverty and hunger, drug trafficking, corruption, gun violence, and the threat of war seem to be on the increase. There appears to be no connection between knowledge and morality but there is a strong connection between knowledge and power and the political influence of wealth. In addition to social problems there is the possibility of extinction in a final nuclear war or a global natural disaster. If the earth's subsystems decide to restore balance in response to abuse, no one, not even the wealthiest, the most powerful, or the most brilliant individuals, will have a place to turn.

Our obsession to explore the unknown and to build gadgets and metropolitan centers—a hallmark of Western culture—that make our lives easier and more enjoyable comes with a price. Some of the crises we face were avoidable; a reversal will depend on worldwide awareness, a spontaneous willingness to change, and whether Earth will tolerate more exploitation.

On these pages I do not focus on the critical issues but on what could have prevented them and what is needed now to change direction and avert disaster. This is not a book about global warming or how to "save the environment." It does not duplicate the work of authors who are academic scholars as well as culture bearers of Indigenous knowledge, especially its ecological aspects. *This book echoes the call of other Natives who have made, and continue to make, valuable contributions to the same goal, utilizing their own knowledge and expertise.* Its content and message is intended to interest and challenge all readers.

Any strategy that purports to be effective in achieving the shift to a new worldview—one that the dominant culture will also embrace—must be convincing. Minimum knowledge and mutual understanding can form an effective bridge that allows communication between cultures. I saw the need for a three-way conduit: a summarized relevant history of Western (or

Eurocentric) science, a historical overview of modern physics that contains the most important discoveries, and a limited but hopefully respectful and accurate representation of Indigenous knowledge.[5] I link relevant information from these areas together.

Most of the information needed for facilitating meaningful dialogues is hopefully included here. The effort is also a gesture of sincerity that implies the dire need for change from the current paradigm. There is much I don't know: the factors and conditions most likely to create the changes we seek; whether or not and the extent to which scientific knowledge[6] is valued by the decision-makers upon whom the changes will mostly depend; whether the information is useful to grassroots movements or to existing efforts to create a global consciousness; and more.

Chapters 4 through 6 are an assessment of our world, primarily from a classical perspective, intended to prepare the reader for subsequent chapters that emphasize the revolutionary discoveries of the 20th century. The contrast is dramatic but mostly noticeable by physicists who are actively working in quantum theory, relativity theory, or M-theory and are looking beyond the narrow expectations of their academic discipline. I try to describe these differences to readers from all backgrounds.

We live in a world of illusions, unaware of unseen realities that support the whole Universe, because we have not become adjusted to the new way of thinking that modern physics implies. Modern physics is not a new path for humanity but it points us to it.

---

[5] This book does not attempt to reveal knowledge which is protected by each tribe. That knowledge is not available to the public.

[6] Unless otherwise indicated, the term "scientific" is a reference to the facts and theories of Western science. The meaning of "science" is discussed in chapter 8.

The vast Universe itself summons us to respond with awe and wonder to the evident harmony, majesty, and intelligence that it demonstrates; as a Native person, I am also conscious of the spiritual dimension that undergirds all things, large and small, near and far. From a physics perspective, there is only one world (that reaches our consciousness); it is the world of quantum and relativity, yet we think in classical terms by habit. But we cannot expect everyone to change if, as physicist Brian Greene tells us, most practicing physicists still don't "feel relativity in their bones."[2]

Because I believe we need lessons in humility, chapter 5 reminds us of the significant and numerous gifts, in terms of inventions and innovations, which Indigenous Americans have made to the world, and other evidences of the enormous knowledge and intelligence possessed by ancient peoples. I decided to include the 15th century Doctrine of Discovery (in the appendix) because it is still in force and because of the lessons it can teach to those who are not familiar with federal Indian law, the notion of "Manifest Destiny," or the question of how a foreign power could legally obtain title to lands that were already occupied. The short answer is "because they could."

Chapters 7 and 8 contain an overview of the quantum story and relativity, as well as a discussion and implications of these theories with my comments interjected along the way. The early 20th century was truly an amazing time of discovery but what most people perhaps don't realize is that much of the work was done collaboratively; several physicists worked in teams and in the same laboratories at different times: the Copenhagen Institute, Cambridge University, Cavendish Laboratory, Göttingen University; experienced physicists mentored those who were new to quantum research and perhaps younger but more brilliant. It was rare when a physicist worked independently; others were involved in other ways besides physical interaction. Unsurprisingly, there were personality conflicts and the rush to publish an idea before anyone else. The fact that some theories were developed independently implies that more than one physicist was working on the same problem; moreover, they pushed themselves hard and, in at least two known cases, a physicist sequestered himself in a remote location in order to work undisturbed until achieving success.

Other topics in these chapters include some of the paradoxes encountered in quantum mechanics and their possible relationship to Indigenous thought, physicist David Bohm's cosmology, and how wholeness—the idea that the Universe is a single undivided whole—is implied by both quantum theory and relativity theory, including the phenomenon of non-locality. These two chapters contain mathematical equations, which are explained for the reader and do not require a knowledge of mathematics.

The term "science" inevitably suffers from different meanings. I have tried in this book to use it consistently as a reference to Western (or Eurocentric) science unless otherwise indicated explicitly or by the context where it is used. Chapters 9 through 12 describe various aspects of science, including the cultural differences that affect its meaning, as well as the effects of colonization on how Indigenous knowledge, or Indigenous science, is taught and practiced in Indigenous communities. Another reason is that the voices of Native scholars like Gregory Cajete of Santa Clara Pueblo and others are getting more attention; his book, *Native Science*, introduces an important dimension to the meaning of science which is lacking in Western thought, one that cannot be expressed in a single English word because the Indigenous concept of knowledge is active and experiential, not static, involving more than the accumulation of facts. For the same reason, Native languages do not have a word for "science." A third aspect is that the meanings of words

The Condor and the Eagle

change with time; prior to being called "science," the study of nature received the name "natural philosophy," which had a broader meaning than it does today, owing to the modern emphasis on empiricism in science.

The history and characteristics of Western science, including the rise of modern science in the 16th century, are discussed in chapters 13 through 15. The importance of this history should not be underestimated. Western civilization traces its origin to ancient Greece, particularly the cities of Ionia along the coast of the Aegean Sea in Asia Minor where relevant philosophical developments seem to have occurred first. What was to become Western civilization took an unfortunate turn when the spiritual teachings of the earliest Greeks were abandoned, creating a split in Greek thought that led to the dualistic philosophy which is characteristic of the West. The people who were destined to become Europeans went astray from their own original instructions; once the path was established, future generations developed a powerful science that helped build societies based on the deeply engrained beliefs of Western culture that violated the fundamental principles of sustainability: take more than is needed; the earth belongs to us; Earth's gifts are primarily "resources" for human consumption, convenience, and profit; nature is "wild" and must be tamed.

The meaning of "the West," or "the Western World," differs among historians but a reasonable view is that the West began after the emperor Diocletian divided the Roman Empire into Greek East and Latin West in AD 285, forming the Eastern and Western Roman Empires, each with its own capital and ruler. Western culture formed from the influences of Graeco-Roman classical and Renaissance culture, Hellenistic philosophy and Christianity, and Central and Eastern European culture. These chapters also explain how the mode of thought shifted from *qualitative*, which characterized the era of natural philosophy, to *quantitative*, which introduced the age of modern science. In other words, modern science represents a new mode of thought from the question of *why* things happen or exist to the question of *how* they can be explained. Modern science is concerned with detailed facts, the experimental method, objectivity, and the synthesis of scientific theory with technology. It is important to avoid confusing *modern science*, which began in the 16th century, with *modern physics*, which began in 1900 with the advent of quantum theory and the subsequent development of relativity theories.

Beginning in chapter 16, the theme turns to actual and potential changes in science as a consequence of recognizing nature's holistic character and the value of a holistic approach. The overall trend is encouraging and hopefully getting stronger. Reductionism can no longer be defended; holism is recognized; academic disciplines are merging or new ones are forming (Earth system science, general systems theory, complex and chaos theory, Green chemistry, etc.) in recognition of their interrelatedness and the holism in nature; green technologies are being developed; some scientists advocate the expansion of the current paradigm in ways that agree with Indigenous concepts. It is not uncommon to discover some aspect of science shifting toward a holistic direction without recognizing the "new" mode as an established Indigenous practice. What is unfortunate is the neglect of such a large amount of information by Native authors that is already available, which is probably due to the public's disinterest in Indigenous issues or vital issues that an Indigenous perspective could help resolve. But in this deeply-divided nation, no one is asking how the nation's Indigenous inhabitants, the American Indian, might be able to help.

What is meant by progress is a matter of perspective: If we want to secure the welfare of future generations, our generation will consciously make decisions to toward that future. But clearly, Western ideology as a basis for sustainable living is a failure; it is an unrealistic and brutal way to treat Earth. Western values contradict the very holism and authoritative principles inherent in nature and confirmed by modern physics. The holistic practice of physics would re-align science into a proper perspective relative to the earth and the entire cosmos while accepting the limitations of the scientific method and extending the domain of knowledge to spiritual and metaphysical realms. *This approach does not advocate a change in physics* but in how physics and all branches of science are applied. The strange behavior of the quantum has been calling for explanations beyond conventional physics.

A holistic perspective of physics has great implications. In the first place, Indigenous people in science can conceive of and practice physics under the assumption that reality encompasses the physical and spiritual domains, which is how I perceive physics. But everyone should at least acknowledge that a re-alignment of science is necessary, that the whole human being—spirit, mind, emotion, body—can enhance the contribution to knowledge. This can be viewed as an effort to repair the consequences resulting from the split that occurred in the 5th century BC among the Greeks.

The second implication is the value of incorporating *both* Native and Western concepts in project design and in the development of science and educational curricula. I recently became aware of two Native individuals who are doing this. Thirdly, the change is necessary to avert the dire consequences of continuing in the current direction. The Indigenous alternative is, and has always been, to embrace the *sacred web of life* and to view the cosmos as a social reality, a web of interconnected and interdependent entities relating to each other to form the whole. Mother Earth is a living being, sustaining us and all our non-human relatives.

Indigenous peoples offer the unique perspective of being in one place for a long time and witnessing the changes around them as they occur. Their stories tell about migrations but they have no memory of immigrating to this land. Old wisdom must re-emerge and be allowed to infuse the modern collective consciousness. Nothing short of a global change in worldview or a shock will bring us to our senses. According to the prophecy, the condor and the eagle once flew together. Of course, we know this; there was no border dividing the north and the south. Knowledge was unified; there was no dualistic split between spirit and matter. The Americas were the homelands of Indigenous nations, given to them by the Creator, and there was no border dividing north and south. Today, we share the land, wealth, and power unequally. When the gifts from Mother Earth begin to disappear, we will be equal again but it will be too late. Now is when we must recognize our interdependency with everything around us; whatever befalls Earth befalls the sons and daughters of Earth. The indisputable message we all need to hear is the following: *Only by saving others can we save ourselves.*

This book is intended for readers with a wide range of knowledge, background, and interests; some topics may be elementary to some readers and challenging to others but it has a message for everyone. Some statements and themes are repeated in different parts of the book but they are made in different contexts. Think of the chapters as separate essays that do not necessarily have to be read in order. I hope this book is encouraging, informative, and of practical value. Specific ideas may be forgotten after reading it but they will hopefully continue to have a positive subliminal effect.

Chapter
1

CENTURIES OF COLONIZATION

For more than two decades, my wife and I have lived near tribal communities where we have met and listened to stories told by respected elders. We have also heard elders speak at conferences and small meetings and read or heard their stories in print or recorded voice or video. The stories are part of a tribe's oral tradition and language; they sharpen listening skills, transmit the history of a tribe, teach moral lessons and respect, clarify concepts, and more.

Stories that share a prophetic teaching also reach my desk occasionally in an e-mail through a Native network. I believe that messages meant to be shared publicly are highly relevant to a divided world of scientific advancements and global conflicts in need of comfort and encouragement—an outcome that is compatible with the purpose of this book and the theme of holistic science.

The tribes and the land have experienced profound changes. The future was not completely unpredictable to many tribal leaders who perceived what military power and the large influx of immigrants could do. How could they not imagine the effects that the greed for land and gold would have on future generations of their people? It was easy to see how the values of so many immigrants, which clashed diametrically with their own, would bring unwelcome change and the end of life as they knew it.

Historians tell us that the stories of non-Western peoples speak of previous worlds, that the world we are now living in is not the first world. The Hopi say that we are living in the fourth world; three worlds were previously destroyed.[3] Our world now faces many crises, apparently more than at any other time, and Indigenous peoples have not been silent about what they see.

The industrial world's practice of planning in short time frames reflects a limited vision about the meaning of history and the future of the world. "We are accustomed to thinking in the time of the human being," says Chief Oren Lyons (Onondaga Nation), "instead of the time of the tree or the rock or the mountain."[4] A 400-year-old tree can be destroyed in ten minutes with no consideration for its value to the forest, the birds or other animals, humans, or the tree itself.

The Prophecy of the Eagle and the Condor

There are several versions of this prophecy but they all have the same essential elements. The condor identifies it as belonging to the Inkas.

This sacred and ancient prophecy is connected with several other prophecies that tell about a future time of healing.[7] It has been preserved by Inkan elders for 500 years since the beginning of European colonization. It speaks of two paths: the path of the eagle, representing the people of the North, who work primarily with the mind, the rational, the intellect, and the path of the condor, representing the people of the South, who work

---

[7] Some of the other prophecies are the Hopi Prophecies, the Return of the White Buffalo, the Return of Quetzalcoatl, , the prophecies of Chilam Balam, the Ojibwe Seven Fires Prophecy, and Black Elk's Daybreak Star Prophesy.

primarily from the heart, intuition, and spirit. It foretold that the two paths would merge during the fourth *pachacuti*,[8] a period of death and destruction, when the eagle would drive the condor almost to extinction and the two would then take different paths. But in the fifth *pachacuti*, the condor will rise again and the two will fly together, wing to wing, as before.

That time which the prophecy foretold is now. We are in the fifth *pachacuti*; the fourth one, which began in the 1490s when the Spanish conquered the Inkas, has ended. The reunion of the eagle and the condor is the time of a great gathering, a reunification of the peoples from the North and South, when they can share their knowledge with each other and the mind and the heart will unite to create a new paradigm, a new worldview in science.

As firm believers in this prophecy, the Otomí peoples built a ceremonial amphitheatre in Mexico dedicated to the "Reunion of the condor [representing the Indigenous peoples of the South] and the eagle [representing the Indigenous peoples of the North]." In 1970, Otomí elders called on their people to build a sacred ceremonial site dedicated to this much-anticipated reunion, rivaling the ancient Mayan and Aztec temples and amphitheaters in scale, artistry, and beauty. The site holds 30,000 people and is adorned with beautiful stone carvings of jaguars and other figures; it has a gigantic carving on top, depicting the reunion of the Condor and the Eagle.

The condor has, indeed, become endangered; it was listed as an endangered species in 1973 and has already been declared extinct in parts of South America. The *Proyecto de Conservación Cóndor Andino* (Andean Condor Conservation Project), one of the groups or agencies that took on the challenge of saving the

---

[8] A *pachacuti* is not a period of *time* but an *era* that lasts about 500 years.

condor from extinction, was created in August 1991 in Argentina.

It is time to apply ancient wisdom and create a healing path that will move the severe social, political, economic, and other divisive issues toward healing, peace, and balance. Societies that separate the sacred from the secular can now set aside their materialistic lifestyle that leads to greed, over-consumption, poverty, and hunger, and make room for generosity, gratitude, fairness, moderation, and justice.

Something must motivate the West to listen to Indigenous teachings, perhaps by recognizing the limitations of its science, noticing how modern physics corroborates several Indigenous principles, considering alternative ways of interpreting physical phenomena, raising the respectability of metaphysics, and taking a comprehensive look at the meaning and history of science. Some Western scientists are, in fact, already advocating a new science paradigm. We must not lose the hope of restoring dignity and honor to Native peoples.

Hegemonic pride can foster doubt about the validity of Indigenous knowledge but no serious physicist will dismiss a concept or interpretation that has a solid basis and strong empirical support. With regard to empiricism, Native peoples are patient and meticulous observers. Still, it may not be easy for a scientist to accept that a scientific discovery which is thought to be new is actually an ancient Indigenous precept that did not require mathematics. Centuries ago they learned and applied principles that were later incorporated into modern science. The concurrence between ancient knowledge and the product of research is qualitative in character and noteworthy because Native peoples, without modern equipment or mathematics, had correctly interpreted what nature was conveying long before those principles were discovered and formulated by modern science. The implications are far from trivial when we consider that some of these discoveries and precepts represent crucial principles of sustainability.

The Legacy of Western Culture

It's not uncommon to hear a Native speaker at a conference assert that quantum physics is now validating Indigenous worldviews; specific examples are not cited, perhaps because they are not easily accessible, but this book provides substantial evidence to support the claim. For over a decade I have been exploring the issue and have found that in several cases Indigenous concepts not only agree with physics but also carry broader implications. Here I describe in detail the specific theories and discoveries that in my view correspond to Indigenous concepts; physics concepts are explained in sufficient detail to allow readers with knowledge and interest to understand the connections and make their own interpretations.

Among the cases I describe (and confirmed by modern physics) is the long-held belief that the Universe is an undivided whole; that all things are interconnected, interdependent, and interrelated—notions conveyed by the often-expressed phrases "all things are connected" and "we are all related" but which carry broader meanings than can be conveyed through equations interpreted from a physics perspective. Within a social context, for example, interconnectedness implies that we (referring not only to humans) *influence* and should *respect* each other. Interdependence is evident by the manner in which humans and nonhumans depend on each other. Interrelatedness speaks of our mutual *obligations* as relatives. If we accept the premise that the proper way to live is linked to a correct understanding of nature, this implies that human societies bear a responsibility to everything within their network of relationships.

But Western societies have not been kind to nature; the inescapable conclusion is that *Western values and practices have, evidently since the beginning of civilization, violated the very principles of wholeness* that nature reveals about itself in order to continue a balanced existence. Since environmentalists now agree that humans are a part of nature, these principles evidently apply to us as well.

Whether these physics principles have spiritual significance is, for many, a question of belief; as for me, I need no more evidence than my own awareness of a Universe endowed with its own consciousness, is intelligent, authoritative, and harmonious—the manifestations of a spiritual reality from which the Creator made all things and to whom we are all responsible. On the other hand, physicists, by and large, have been quick to warn against alleged parallels or other allusions to "religion." This response is not surprising; it demonstrates how modern physics, or shall we say nature, is suggesting mystical ideas on its own with no help from outside of science.

The inference about spiritual significance is reasonable also because Indigenous people contemplate not only the quantitatively measurable behavior of a phenomenon but also its *character*, considering how all things in the cosmos have the possibility of relating to each other, a relationship that includes the observer. Thus it is likely that, compared to an impersonal approach to research, living in a respectful and mutual relationship with nature will yield results that carry a deeper significance. Nature's strange behavior is puzzling but I'm convinced that, since we are only one "strand" in the sacred web of life, research and discovery carry a human obligation. Some phenomena may be incompletely knowable or unexplainable and remain part of the great mystery.

For minds committed to empiricism, physics may have little meaning beyond what the physical realm reveals; it does not understand the obligations of reciprocal relationships because of its fragmented and incomplete view of reality. It divides knowledge into categories that focus on particular aspects in isolation, so that neither the whole nor the relationships to other aspects are considered together. Specialists working in the same or other fields will further refine that fragmented view. As an example, consider the vitamin and mineral supplements which are prepared and sold separately, based on the knowledge that our bodies need them for certain bodily functions, without considering whether they work most effectively when foods are consumed together as part of our diet. Or, consider the possible and known negative health repercussions of growth hormones given to animals from which food products, or the animals themselves, are consumed by humans.

It is safe to say that profit is a factor in many of the decisions that ultimately affect us as well as the animals which are exploited. In contrast, a holistic perspective takes into account *all factors* which may be present, so that no parties will have an unfair advantage. This expectation is unrealistic unless a change in worldview is also advocated.

What legacy is a culture likely to leave if its values have been working against nature for centuries, even millennia? *The science of Western civilization has from the beginning separated humans from the natural world,* creating a hostile, rather than a loving, relationship. Modern society views the earth's gifts as commodities and resources; it is a culture of over-consumption, hostility, conquest, and greed, driven by deeply engrained values: take more than you need; the earth belongs to us; tame the wilderness; consume the "resources"; make a profit. The environmental impact of *political* and *economic* crises is evident. The richness of the soil, the purity of the waters, the clean air that our ancestors enjoyed: all has changed; it is the inevitable outcome of a unilateral and hostile relationship toward the earth. The national debt and deficit in the United States are each measured in terms of trillions of dollars. The nation is also involved in international armed conflicts. Corporate greed is winning; the symbolism and fervor of Manifest Destiny—industrial development, institutions, metropolitan centers, economic prosperity, cutting up the land, etc.—with all of the promises that reflect Western values, are unrelenting obsessions.

Yet in science and technology, the achievements of Western civilization during my own lifetime alone have been dramatic. But how do we judge progress? During my childhood in the early 1940s, our family had our first telephone installed, followed by television in 1950, and later we witnessed an amazing growth in information technology: transistors, computers, automation, space exploration, and more. In the late 1950s, the first-generation computer I worked with was as large as an upright refrigerator; it had a drum memory with a capacity of less than three thousand 29-bit words and had 27 miles of wiring, 580 vacuum tubes, no programming language (much less a GUI—graphical user interface), and required its own refrigerated air conditioning. Modern technology is great but we're paying for it from our grandchildren's account.

We are also witnessing the desecration of sacred places, the invasion and destruction of Indigenous lands and social systems, and a warmer climate, not to mention the social ills that affect entire populations: violent crime, drugs, wars, poverty, and so on. Can we attribute all of this behavior to the human condition? Or will we, finally, heed the teachings our ancestors and elders that the earth is our mother, that all things are connected, and that whatever we do to the land, we do to ourselves.

The obsession with exploring the unknown just "because it's there," industrialization that makes life easier and more enjoyable due to ever-improving technologies, and metropolitan centers that build or import all the goods we need and want, come with a price. Some of the crises we now face were avoidable and may be irreversible. Reversing the process of social decay will depend on worldwide awareness, more than a spontaneous willingness to change, and whether the earth will tolerate more abuse. Old wisdom must re-emerge and be allowed to enter into the modern consciousness—a wisdom that goes back to the earliest times in causal history when important sustainable principles were abandoned. Nothing short of a global change in worldview or a widespread shock will bring us to our senses. *Only by saving others can we save ourselves.*

This past decade alone witnessed a rise in the significance of globalization in all its aspects: communication, economic status, war, and natural disasters. Due to global dependencies between American and European banks, the economic meltdown experienced by the U.S. in 2008 was felt internationally.

Calling it progress, implying that the world is getting better and benefiting everyone, including all living things, is not an accurate assessment. How we perceive change, though, depends largely on our worldview. Eventually, we must ask about the survival potential of species and the earth's natural systems and whether the world's powerful governments are prepared and serious about reversing the direction and trends. Meanwhile, the natural world continues at its own pace in a state of flux, not unaware of or unaffected by human activity.

Indigenous peoples around the world saw a global crisis coming and issued warnings but the industrial world did not listen; it continued to sleep until, for them, it came unexpectedly. Long ago, European inhabitants went astray from the original instructions and future generations followed the path that was established, eventually developing a powerful science that helped build societies in accordance with commonly-held beliefs which they tacitly took for granted.

In the landmark case, *Citizens United v. Federal Election Commission* (2010), the U.S. Supreme Court held in a narrow 5-4 decision that the First Amendment prohibited the government from restricting independent political expenditures by corporations and unions. This means that these entities are free to spend unlimited amounts of money lobbying for or against candidates who are running for office. And according to news reports, when oil company executives came before the U.S. Congress in 2008 to answer questions about their exorbitant profits, their reply was "it's no big deal"; in fact, one executive defiantly suggested that closing big oil's tax loopholes is somehow "un-American."

Moral decay is only one indicator that the Western World is out of proper relationship with the earth. The Ojibwe Prophecy of the Seven Fires foretold the arrival of a light-skinned people from the East who would be given a choice between two roads. If they choose the wrong one, "the destruction they brought with them will come back to them." But if they choose the right road, the seventh fire will light the eighth and final fire of peace, love, and brotherhood.[5,9] Has the nation reached a point of no return?

A Science that Everyone Needs

Before European contact, American Indian tribes lived sustainably on their lands. Those ways were, and still are, part of a tribe's spirituality, which is connected to the land. The people, their knowledge, the land, and their spirituality are one system—a land ethic based on the principle that humans are

---

[9] Indigenous prophecies do not predict the future; they often warn of consequences if certain conditions are not met, such as violating a natural law. According to Mayan elders, the 2012 prophecy did not predict the end of the world.

part of the creation, not separate from it or surrounded by it. It was only recently that the term "environment" began to give way to this ancient principle as if it were a new idea. The natural world tries to maintain balance, order, and harmony; it teaches us to maintain the same pattern of living in reciprocal relationship to all things with which we interact. The first question we should ask when encountering a plant, animal, or mountain is, What is my responsibility to it?

Thousands of Indigenous nations worldwide far outnumber the nation-states which were formed from the conquest of tribal peoples. Wherever we walk in America, we are likely to arrive upon a tribe's original homeland. When we come upon a place with the right attitude, we can learn the meaning of sacredness.

The tribes and the land have experienced profound changes—a future that many tribal leaders perceived when they witnessed the potential of military power and foresaw the large influx of immigrants. How could they not imagine the effects that greed for land and gold would have on future generations of their people? It was easy to see how immigrant values, which clashed diametrically with their own, would bring unwelcome change and the end of life as they knew it.

Some tribal leaders also had prophetic insight. In the 1850s, during a terrible period in this country known as the "Indian wars," Sweet Medicine of the Cheyenne told his people:

Soon you will find among you a people who have hair all over their faces, and whose skin is white. They will be looking for a certain stone; they will be people who do not get tired... These people will not listen to what you say; what they are going to do they will do. They will try to change you from your way of living to theirs. They will tear up the earth, and at last you will do it with them.

When you do, you will become crazy, and will forget all that I am teaching you.[6]

The United States did what it was going to do. As the settlers moved across the land, American Indian tribes were subjected to unacknowledged genocide. Indians north of the Rio Grande who survived were subjected to various efforts to assimilate them into the immigrant culture with the intent to eradicate all forms of Indian life. The holocaust that occurred throughout the American continent due to the spread of diseases and wars of conquest extended as far south as Tierra del Fuego.

The teachings of which Sweet Medicine spoke, and more generally those of the world's Indigenous peoples, reflect detailed knowledge of their environments and the relationships between all things within their complex ecosystems. Indigenous knowledge is inseparable from the Native people who possess it; it comes from an empirical and experiential way of life that involves the heart, the head, and the spirit.

The knowledge and perspectives that guided the First Peoples were scorned in favor of a European worldview that is unsustainable and disrespectful because it views Earth as a commodity and a bottomless resource, violating the reciprocal relationship that must exist between human and non-human. We are now dealing with the consequences of a madness that began centuries ago. The technologies developed in North America were never intended for the welfare of the First Nations. Westward expansion was fueled by greed and a vision about a new world that did not include the Indian. Perceiving Indians as "ignorant savages," the immigrants used discoveries in science to build technologies and new industrialized societies to benefit themselves. Their children know little about the Indian side of American history, Native thought, or ancient wisdom—a knowledge gap that they carried with them into adulthood and into the influential positions that they now occupy within the hierarchical structures of society. Now we all face the harsh realities that may require a return to simpler lifestyles and fewer conveniences. We may have to learn the hard way to show respect for Mother Earth in order to survive.

In a symposium presentation at the 2003 annual meeting of the American Association for the Advancement of Science (AAAS), in which I participated on a panel, Pueblo author and educator Gregory Cajete, also on the panel, stated "Indigenous science is not just for Indian people; everyone needs it to address the problems of this century." A respectful relationship with the earth is an essential tenet of Indigenous wisdom. Our relationships to Creator, to ourselves, to the natural world as stewards, and to other human beings and nations are deteriorating. Restoring balance, if it is still possible, will require a revolutionary change in thought and practice. In 1680, when the Pueblo Indians of New Mexico revolted against the Spaniards as a last resort due to almost one hundred years of abuse, Pueblo leaders said, "The world is out of balance."

America's patriotic songs speak of the birth of a nation and the possession of a homeland. Americans have appropriated several principles and symbols of democracy from the tribes. One notable example is the symbol of an eagle holding a bundle of thirteen arrows in the seal of the United States, which was co-opted from the bundle of *five* arrows as a symbol used by the Haudensosaunee, or the Iroquois Six Nations Confederacy. The bundle, of course, symbolizes the strength of multiple nations united together compared to that of a single nation. The watchful eagle of the Haudensosaunee is placed at the top of the Tree of Long Leaves in the Great Law of Peace. Also, as Onondaga Chief Oren Lyons once pointed out, the founders omitted the Creator from the U.S. Constitution and the new American nation also embraced slavery. These were signs, says

Chief Lyons, that America's flame of democracy was already beginning to diminish.[7]

America and other industrialized nations possess the technology of peace and the technology of war but even the instruments of peace can be destructive when the sacred is disregarded. America's own path has always involved choices. Traditional Indigenous elders tell us that the problems have spiritual roots and that humanity must awaken to the urgent need to care for Mother Earth and all life. When the technologies of war and the mindset of conquest were used to appropriate the territories of Indian nations that now reside within its borders, America became a nation-state. But her claim to nationhood implies that she has the soul of a nation and is responsible for the choices she makes.

A Forgetful Nation

"Nevertheless, Greece was the mother of Europe, and it is to Greece that we must look in order to find the origin of our modern ideas."[8] Thus spoke Alfred North Whitehead, eminent mathematician and philosopher of the early 20th century, comparing the progress of scientific knowledge in Europe in the 17th century to that of ancient Greece.

Western culture is proud of its Greek roots and, it is fair to say, so is modern conservative Christianity.[9] But it is a heritage that was taken for granted because there is no visible evidence, at least not in America, that people today celebrate the *spiritual* origin of Western culture. That's right: European Americans and Europeans can trace their genealogical or national roots to Greece but have no concept of the sacred roots of their ancestry; these are the roots that were abandoned in the development of Western civilization.

The memory of those roots was lost to the West long ago, according to Peter Kingsley, the author of several books and internationally recognized for his original work in ancient Greek philosophy and culture. Kingsley explains that, like Native Americans, the West has ancestors of its own who gave them original instructions; these early Greek masters of wisdom who preceded Socrates —Pythagoras, Parmenides, Empedocles—laid down the spiritual laws for Western civilization as it was originally meant to be. Kingsley discovered that, although the West lost track of those highly significant spiritual teachings, the medieval Persian Sufis and alchemists who are outside of Western culture had preserved and taught the essence of what the West forgot.[10]

The discovery of something as precious as these teachings—something that was lost or not known to have existed—can be an ecstatic moment and the beginning of a healing process that may lead to more discoveries. I know this from my own experience and have also known that others have had similar experiences but I do not know whether a collective consciousness of healing is possible among a populace within a nation.

A shared consciousness can, of course, take the form of trauma instead of ecstasy. If we ask whether the members of the majority have experienced shock or shame after learning that atrocities were committed by their nation against a minority population, we immediately think of Germany and the Jewish Holocaust. The question of why there is no American consciousness about the Indigenous holocaust that occurred on this continent still awaits a response. But it is difficult for me to accept that there will be no long-term consequences. It seems to me (and perhaps many others) that many Americans are spiritually lost, psychologically unbalanced, and confused, unaware of what is lacking in their knowledge of causal history but are suspicious, nonetheless, that something horrible must have occurred and are afraid of what they would learn by examining that part of history. I was once one of the many church-going Christians who isolate themselves from information sources that contradict our deeply-seated beliefs and focused on literature and news stories that support our views.

As a way of better understanding multi-generational trauma, it may help to consider the analogous experiences of Indian children, now adults, who were adopted out of their parents' homes, lived in an alien (American) culture during their growing-up years, and as a consequence of losing their identity, now suffer psychological trauma. Some of them were fortunate to find help through programs like Gene Thin Elk's Red Road Approach, White Bison's Wellbriety Movement, the Sacred Tree Curriculum of Four Worlds Institute, or other programs which employ culturally relevant methods, like the sweat lodge, to help individuals re-discover their identity as the first step in the healing process.

My point is that there are reasons why America seems to have no conscience about the American Indian experience. The tribes were sovereign nations long before the United States could consider its own sovereignty. The Iroquois Confederacy was the first sovereign entity that granted recognition to the new American nation comprised of confederated colonies under a central government. The United States needed this recognition among the nations on Earth and by denying her role in the history of the American Indian, she forgets how the nation was born, choosing to remember and perpetuate only the bravery of her own soldiers in the cause of freedom.

It would be highly appropriate—indeed, an obligation—for America to memorialize the countless Indian warriors who also sacrificed their lives. The tribes have their own memorial ceremonies, which my wife and I have attended, for warriors who fell as they defended the cause of freedom, first against and later for the United States. American Indians represent the highest per-capita commitment of any ethnic population to defend the United States.[11]

When we learn about a massacre, even in time of war, the shocking news awakens our consciousness and we often wonder how people could commit such atrocities. Something is wrong if we ignore it and behave as if nothing had happened. American journalist Geoffrey Wolff wrote a statement that appeared on the inside back cover of one of the numerous printings of Dee Brown's classic book, *Bury My Heart at Wounded Knee*, first published in 1970, which is a meticulously documented account of the American Indian's near extermination during the last half of the 19th century—a period in U.S. history known as the "Indian Wars." After reading the book, he said: "It falls to a journalist reviewing the books of our days to treat the dreadful almost as though it were commonplace. The books I review, week upon week, report the destruction of the land or the air; they detail the perversion of justice; they reveal national stupidities. None of them—not one—has saddened me and shamed me as this book has. Because the experience of reading it has made me realize for once and all that we [Americans] really don't know who we are, or where we came from, or what we have done, or why."

Michael Moore, an American filmmaker, author, social critic, and activist, was deeply disturbed by the massacre of twenty school children and six staff members at Sandy Hook Elementary School in Connecticut in December 2012. He asked why such gun violence happens in this country. "They [other countries] simply don't kill each other at the rate that we do. Why is that?" he asks, then continues: "This probably shouldn't come as a surprise to us as we are a nation founded on genocide and built on the backs of slaves. We slaughtered 600,000 of each other in a civil war. We 'tamed the Wild West with a six-shooter,' and we rape and beat and kill our women without mercy and at a staggering rate: every three hours a women [*sic*] is murdered in the USA (half the time by an ex or a current); every three minutes a woman is raped in the USA; and every 15 seconds a woman is beaten in the USA... We think we're #1 in everything when the truth is our students are 17th in science and 25th in math, and we're 35th in life expectancy. We believe we have the greatest democracy but we have the lowest voting turnout of any western democracy. We're biggest and the bestest at everything and we demand and take what we want."[12]

Today, more than one hundred years since the end of the Indian wars and forty years after Brown's book began to stir the consciences of millions of readers, it is appropriate to ask whether there have been any improvements in the status of the American Indian or changes in the nation's conscience. One would think that since the Indian wars are over the tribes have nothing more to fear from Americans or their government. Sadly, the forces working against the Indian have not relented. The "Indian wars" are no longer fought on the battlefield but in the courts. The list of socio-economic ills, ongoing political conflicts, government and societal abuses, historical trauma, and other issues is too lengthy and complex to include in this book; we would have to deviate from our main purpose.

However, one recent incident reported in the national news does deserve mentioning because it exemplifies the attitude, ignorance, and heartlessness that still exists *today* among members of the U.S. Congress. In January 2012, U.S. Senator Rand Paul (R-KY) introduced two budget-reduction ideas that essentially requested Congress to ignore all treaty obligations to the tribes, conveying as clearly as possible his low regard for tribal communities. His proposal would (1) eliminate the Bureau of Indian Affairs (which would also eliminate all federal Indian programs) and (2) cut in half the Department of Health and

Human Services' Indian Health Service. Had Congress adopted those ideas—and it has the power to do it—it would be the end of Indian reservations and much, much more; the results would be disastrous. The United States is obligated to uphold its treaty obligations to the tribes; treaties are the Supreme Law of the Land, equal to the U.S. Constitution, and even bind the states (Article VI), yet tribal issues do not rate as sufficiently important to be placed on the national agenda, even during presidential campaigns when candidates address the critical issues of the nation.

The senator needs to hear what Australian Prime Minister Paul Keating said during that country's 1993 celebration of the International Year of the World's Indigenous Peoples: "It might help if we non-aboriginal Australians imagined ourselves dispossessed of the land we lived on for 50,000 years, and then imagined ourselves told that it had never been ours."[13] The tribes in the United States have been on this land for at least that long.

It is standard practice for American historians to include *Manifest Destiny* in their accounts—the belief and the sense of mission that America's westward expansion, which meant taking over the continent, establishing institutions, and seizing Indian lands, was divinely ordained. Think of what this implies, especially considering the significant number of Christians who at the time were also committed to this doctrine: that *God approved* the elimination of 95 to 98 percent of the Indian population and everything else that belief in the doctrine implies! There is no justification. But when people are strongly committed to an ideal, they search for evidences that seem to support that ideal, such as comparing what happened to the American Indian to Israel's invasion of the Canaanites in the Old Testament.

After the last Indian massacre in 1890 at Wounded Knee in South Dakota, Americans believed that the conquest was complete, by then claiming vast territories while demoting the original inhabitants to an inferior status—a perception bolstered by Hollywood films that are still being shown.

Americans need to understand their own lost condition before considering themselves divinely-justified conquerors. Until then, both major political parties, apparently oblivious to these questions of conscience, will continue to claim that America is theirs either because they came through a (legal) immigration process or by "right" of conquest. When one party challenges another and each party claims that it needs to "take back our country," it is clear that they are still lost and confused. Since it is often said that we must not rely on emotions but on facts, consider that "legal" immigration disregards the historical fact of numerous land acquisitions which occurred through deceit or conquest.

On what basis is immigration legal if the land was acquired by theft, deceit, or conquest? Given the fact that so much time has passed but the descendents of the original land owners or occupiers are still living, what are the nation's and citizens' moral and legal obligations to them? What Christian teachings, principles, or values are applicable? How should citizens who claim that America was founded on Christian principles respond to these questions which involve matters of conscience and Christian values?

Returning to Empedocles and the early Greeks, Empedocles' philosophy is significant not only because it contains the foundational principles of Western civilization but also because it agrees with the Indigenous principle of connectedness and interdependency. According to Presocratics scholar Patricia Curd, Empedocles believed that the proper way to live is tied to a correct understanding of nature. She writes:

The Condor and the Eagle

"The correct philosophical understanding of the physical world and the correct way to live cannot be separated from one another in Empedocles' thought (a similar attitude appears in Heraclitus); one cannot fully understand the world without living correctly."[14]

But as you continue reading this book, the converse of Empedocles' belief will also hopefully become obvious, which is that nature has taught us many holistic principles (through modern physics and in other ways) which imply human responsibility. Then again, this is a crucial point: when nature speaks, we need to listen and properly interpret physical phenomena with the light that Spirit gives to our minds; the interpretation is subjective and the relationship is mutual.

American Indian Education

Science education should be attracting more Native students, in view of the relationship that their ancestors must have sustained with their local environments during a time when virtually all tribes practiced traditions that are rooted in Mother Earth—as many are still doing. Since science is about nature, Native students' inherited connection to nature should motivate them to study science. In view of this connection, science education needs to be culturally relevant. The obvious lack of interest in science is due, at least in part, to the lack of relevance.

But there are other factors, not the least of which is the influence of Western values that are ever present and difficult to avoid, depending on each person's circumstances. The educational system, including college-level education, television programs viewed in the home, friendships, and the various ways in which Western society separates of humans from nature are all influences. As an example, consider how small children enjoy films that involve fantasy or have animal characters. Do they have realistic nature scenes? Do the animal characters act like people or like intelligent animals? Do they refer to animals as "Mr. Bear" or simply "Bear"? Do the roles that family members play reflect the nuclear family or the extended family? Are natural phenomena portrayed as fantasy or as real events? Do the films have spiritual dimension?

Still more factors of influence are: an irrelevant classroom environment; the textbook(s) used; the teacher's familiarity with tribal cultures and worldviews; whether teachers acknowledge the students' own knowledge; the lack of knowledge of science areas that relate to nature. There is an obvious need for teachers and college faculty who can incorporate relevant tribal knowledge and perspectives with the teaching of science.

Students need to be empowered in accordance with their ancestral heritage so that the multi-generational relationship with the natural world can continue. Properly understood, one can see how a tribally relevant science education can bolster the human potential for survival, a matter of universal concern, not one for scientists alone to address.

Relating Indigenous knowledge and perspectives to authentic physics can have a positive global impact in Western circles. If physicists and others in science can be convinced that they can embrace the web of life—a unifying principle that implies universal responsibility—without compromising science or involving religion, they can help raise the consciousness of humanity to a global scale.

Science textbooks are problematic, though, if they are not free of Western cultural bias. One physics text that I used in a course echoes Hans Eysenck's dictum: "If it cannot be measured, it does not exist." It tells students what is animate and what is not, denigrates tribal cultures as "primitive," and uses references and innuendos that reflect the author's own culture and beliefs that are only remotely related to physics but are included, no doubt, because of the author's hang-ups with "pseudoscience." The comments made by authors and the teaching approach give students from the non-Indian world the mistaken idea that Indian cultures have nothing of value to contribute to science.

Science education transmits the values of a community of scientists. When Western cultural content (not scientific fact) conflicts with tribal values, it adds to other factors that discourage tribal students from entering science careers. The attitude behind cultural biases such as those expressed above is now being challenged, according to Glen Aikenhead, one of the scholars who began in 1975 to "demythologize" the scientific fundamentalism that exalts Western science.[15] Aikenhead, professor emeritus of education at the University of Saskatchewan, has written extensively about aboriginal science education in Canada.

On a wider scale, public Indian education in the United States has gone through various stages of government actions and failures. Prior to contact with Europeans, the tribes on this Continent had their own methods of educating individuals at different stages in their lives. American Indian tribes, like tribes in other regions of the world, have had to adapt to profound changes in order to survive. In the latter part of the 19th century, it was once widely believed that the extinction of the American Indian was only a matter of time; in fact, newspaper editors called for their extinction. But at the end of that period, when the United States had achieved its goal of reaching the Pacific coast, the Indian had not disappeared after all, albeit the Indian population declined precipitously to a low of about 250,000 in 1900 (reflecting a depopulation rate of ninety-five percent or more). The last Indian massacre occurred at Wounded Knee, South Dakota in December 1890, essentially bringing an end to the American frontier.

Armed conflicts ended but a new problem now confronted the U.S. government and American society: what to do with the tribes that survived. Many had lost their lands and would no longer be treated as enemies but they suddenly had to cope with profound changes, representing a new kind of struggle for survival. Some tribes were forcibly moved to reservations, an action that did not completely solve the "Indian problem." In an effort to absorb Indians into American life, the United States established a national policy of assimilation through boarding schools (similar to the residential schools in Canada) that were designed to make them forget everything they knew about being Indian and to transform them into "useful" American citizens; it was a strategy to eradicate all traces of Indian culture and language and to teach them the American way of life within the context of Christianity. The boarding schools were actually *immersion* schools operated by Christian churches and mission groups—a national policy that blatantly violated the First Amendment and undoubtedly constituted cultural genocide. The boarding school era lasted almost one hundred years.

During a brief era of positive change which began circa 1933, Commissioner of Indian Affairs John Collier questioned the materialism of modern American society and embraced the value of traditional Indian religion. Unfortunately, toward the end of World War II the government, influenced again by conservative politics, tried to terminate Indian reservations and achieve the complete cultural assimilation of Indians. Several reservations *were* in fact terminated, causing many Indians to relocate permanently to the cities on the (false) assumption that jobs would be available—an action that created new problems associated with urban life. In 1969 a Senate subcommittee report (*Indian Education: A National Tragedy, a National Challenge*) documented the continued failures of Indian education. Recognizing the Indians' aspiration for self-government, the Nixon administration founded the Navajo Community College (now Diné College), the first tribal college, in 1968. Tribal self-determination was further advanced with the passage of the Indian Education Act in 1972 and the Indian Self-Determination and Education Assistance Act in 1975.

Despite these positive changes, the high school dropout rate among Native students (36%) was the highest of any "minority" group in the United States, according to the U.S. Secretary of Education's Indian Nations at Risk Task Force, which held testimonial hearings in 1990 and 1991. They identified several factors that contributed to the dropout problem: an unfriendly school climate that failed to promote appropriate academic, social, cultural, and spiritual development, a Eurocentric curriculum, low teacher expectations, lack of Native educators as role models, and overt and subtle racism. The final reports states, "The renewal of traditional Native cultures in and out of school is re-establishing a sense of community and is fighting the materialistic, hedonistic, and individualistic forces of the popular culture"[16] — an assertion that unmistakably condemns a culture that is detrimental to Indians. On the positive side, the Task Force found that schools that supported a student's language and culture were much more successful, and in 1993, the United Nations recognized the predicaments and aspirations of Native communities and called for a renewal of traditional Native cultures.

It should not be difficult to recognize that Indian students are affected by factors that do not apply to other students. This is not a trivial matter; Native psychologist Eduardo Duran has coined the term *soul wound* with reference to the widespread historical trauma in Indian country, including a high incidence of post-traumatic stress disorder, which is a multi-generational condition.

Culture plays a significant role in public education. Much of the information we absorb daily is conveyed through the culture of the dominant society. The values of that culture also influence us through the news media, television programming, church, and other ways. A society's language also reflects its culture, and this language has become dominant in many homes where Native languages are also used. These influences can subliminally affect an individual's foundational philosophy, or worldview, which operates tacitly without prompting when a judgment or decision is needed. Within this worldview, we contemplate the world, interpret experiences, and assert what is real without thinking much about assumptions that have become deeply seated.

The Elders Speak

The elders' voices may not be audible but they are not silent. The International Council of Thirteen Indigenous Grandmothers, formed in 2004, represent virtually every region of the world: the Amazon rain forest, the Arctic Circle, the forests of the American Northwest, the plains of North America, the highlands of Central America, the Black Hills of South Dakota, the mountains of Oaxaca, the desert of the American Southwest, the mountains of Tibet, and the rain forest of central Africa. The Council was formed out of concern for the "unprecedented destruction of our Mother Earth," which they cite specifically: contamination, war, poverty, the threat of nuclear weapons and waste, the prevailing culture of materialism, health epidemics, exploitation of Indigenous medicines, and the destruction of the Indigenous ways of life. The following paragraph is a part of their statement:

> We, the International Council of Thirteen Indigenous Grandmothers, believe that ancestral ways of prayer, peacemaking, and healing are vitally needed today. We come together to nurture, educate, and train our children;

uphold the practice of our ceremonies, affirm the right to use our plant medicines free of legal restriction, protect the lands where our peoples live and upon which our cultures depend, safeguard the collective heritage of traditional medicines, and defend the Earth Herself. We believe that the teachings of our ancestors will light our way through an uncertain future.[17]

Several Indigenous people from various regions of the earth customarily monitor the climate through direct observations that they sustain throughout the year. The Inuit in Canada's High Arctic, who conduct extensive research on climate change, have been reporting that autumn freeze-up is occurring a month later.[10]

The Indigenous Kogi people, survivors of a pre-Columbian civilization, live in a hidden place high up on Colombia's Sierra de Santa Marta in South America, which they call "the heart of the world." They believe they are the guardians of all life on Earth and call themselves the "Elder Brothers" of humanity. They have been observing climate changes and became so concerned that in 1998 they invited Alan Ereira, an award-winning British radio and television producer, to visit them in order to convey their message on film, which is available in a documentary video as well as in paperback.[18] The Kogi are telling their younger brothers (Western civilization, people from the industrialized world) that "Unless we work together, the world will die," because they are "killing the earth."

The Inuit in Canada's High Arctic have been reporting warmer temperatures and changes in animal behavior that portend long-term climate change. In 2000, they produced a broadcast-quality video, "Inuit Observations on Climate Change," which follows the local people on the land and sea as they perform traditional activities.[19] Alaska Natives have also

---

[10] Visit www.iisd.org/casl/projects/inuitobs.htm

been warning about the warmer temperatures and changes occurring, for example, in animal behavior. Their extensive educational resources are available through the Alaska Native Knowledge Network.[20]

At the historic *Cry of the Earth* conference [11] held near the end of the United Nations International Year of Indigenous Peoples (November 1993), delegations from seven Indigenous nations in the Western Hemisphere—Hopi, Huichol, Maya, Iroquois, Algonquian, Mi'kmaw, and Lakota—delivered their prophecies in person to the United Nations. The prophecies were translated into English as each delegation spoke in its own language. This is an important resource. Videos of the event may be purchased, borrowed, or viewed online.[12]

A significant milestone was reached in September 2007 when the United Nations adopted the Declaration on the Rights of Indigenous Peoples after thirty years of debate. The struggle began in 1977, when the Haudenosaunee (Six Nations Confederacy, or Iroquois) delivered a major address to the Western World in the form of three position papers, titled "*A Basic Call to Consciousness: The Haudenosaunee Address to the Western World.*

This was not the first time that the Haudenosaunee had presented a grievance before the United Nations. In 1923, Deskaheh, Chief of the Younger Bear Clan of the Cayuga Nation, traveled to Geneva to bring their case before the Council of the League of Nations (replaced in 1945 by the United Nations).

---

[11] A recording that contains the entire conference may be viewed online and on video; just do an Internet search on "Cry of the Earth Conference."

[12] The videos, which were produced by Crescentera, may be purchased from the Wittenberg Center (http://www.wittenbergcenter.org/id5.html). They may also be viewed online or read in English by using the search phrase "cry of the earth conference." They are also available from several university libraries.

In their 1977 address, the Haudenosaunee state the following:[21]

> "The air is foul, the waters poisoned, the trees dying, the animals are disappearing. We think even the systems of weather are changing. Our ancient teaching warned us that if Man interfered with the Natural Laws, these things would come to be. Our way of life is disappearing. ... The West is burdened by the weight of centuries of racism, sexism, and ignorance. ... We must all consciously and continuously challenge every model, every program, and every process that the West tries to force upon us."

Basic Call is an insightful and incisive examination of Western civilization, including its historical development. It represents a call to consciousness of the sacred web of life in the Universe. They begin with early history, before the Indo-Europeans came to Europe or North America. On more than one occasion, I have heard an elder admonish the people against attempts to change how nature functions or finding ways to circumvent nature. But Western civilization has done just that, as in the following cases which *Basic Call* points out:

Around ten thousand years ago, peoples who spoke Indo-European languages lived in the area known today as the Steppes of Russia, which stretch from the Ukraine eastwards to the regions north of the Black Sea and the Caspian Sea. They had developed agriculture, and it is said that they had begun the practice of animal domestication. This involves changes at the genetic level that are arbitrarily achieved by humans for the purpose of creating or amplifying characteristics that benefit humans. Domestication changes the *species*, whereas taming causes animals *individually* to adjust to the presence of humans.

The hunters and gatherers who roamed the area probably acquired animals from the agricultural people and adopted an economy based on the herding and breeding of animals. At that time, they were a Natural World people who lived off the land.

Much of the history of the ancient world recounts the struggles between the Indo-Europeans and the Semitic peoples. Over a period of several millennia, the two cultures clashed and blended; by 2000 BC, some of the Indo-Europeans, the Greeks, adopted the practice of building cities.

The first Indo-European tribes in Europe to spread into other parts of the world and make significant progress were the *Greeks*, who moved south into Greece and the Aegean from the 18th century BC. The herding and breeding of animals signaled a basic alteration in the relationship of humans to other life forms. Previously, humans depended on nature for the reproductive powers of the animal world; they assumed the functions that for a long time had been the functions of the spirit of the animals. The creation of towns and cities implied a new need for imports and exports and water. Finding ways to bring water (a form of spiritual life) to the cities was another way in which humans replaced the function of nature. This led to supportive technologies, hierarchical organizational structures, concentrated populations, and exploitation of the natural world. Importing from the countryside implied exploitation of the land; smelted iron and other tools by Indo-Europeans were used for agriculture to cut down trees and other functions, causing workers to be displaced, which led to conflicts. The rise of the European state meant expansion and war, world conquests, the need for fuel, and further exploitation of the natural world.

In summary, the development of Western civilization involved the creation of technologies that modified or substituted for nature's ways. How we relate to nature accounts for the profound differences between Western and other technologies.

Chapter
2

CHANGING THE PARADIGM

Colonizing interests can be expected for various reasons to resist any attempts to change the current science paradigm. They may feel, for example, that the current emphasis on empiricism is somehow being threatened or that science will be contaminated by religious or superstitious ideas. Whatever may be the reaction from the science community, it is encouraging that some Western scientists, who are also aligned with the need for change and are focusing on science's deficiencies, are taking a broad-minded attitude and propose re-defining science by expanding or replacing the current paradigm. It is further encouraging that, as I will show, the reasons they give are in agreement, at least to a degree, with the tenets and values that Natives want to preserve, although ironically, the sustainable concepts which they now embrace are actually tenets that Native peoples have long known and practiced—a fact that is not acknowledged. But in no case is it necessary or even suggested that authentic empirical methods be abandoned or that we should not distinguish between opinion and fact.

A strategy that claims to be effective in achieving the shift to a new worldview—one that the dominant culture will adopt—must be convincing. The issues that need to be overcome are not trivial but I'm convinced that an effective approach will include efforts to achieve a certain level of literacy among all participants in the concepts of modern physics, Indigenous knowledge, and Western science—the areas covered in this book. Minimum knowledge in these areas and mutual understanding can form an effective bridge that allows communication between cultures. They can also be a gesture of sincerity implying the need for change.

This book contains an adequate amount of information for readers who possess the proper level of maturity to achieve basic literacy in (a) the concepts of modern physics and Indigenous knowledge and (b) the history and characteristics which form the Western science paradigm. As to what is adequate, proper, or basic, these are determined by the approach itself; by telling the story of how quantum theory developed, I introduce concepts that require explanation and, as these concepts are described, I comment on how in my opinion they relate to Indigenous concepts and knowledge. Regarding "knowledge," only the concepts can be described to an extent but nothing can replace experience. The history of Western science spans more than two thousand years, during which some thought patterns and philosophies changed while others endured to form the current paradigm.

Scientific facts may not be the primary factor that will create the changes we seek because hierarchical structures in our society often give decision-making responsibilities to less-knowledgeable people; too often good ideas are stifled and priorities are changed by people with a higher authority. Institutional problems like this will somehow have to be overcome so that good ideas can be heard.

Science with Spirit

Tribal institutions and educators, particularly administrators and other decision-makers who bear a large measure of responsibility for curriculum planning and also have influence, will hopefully utilize this book to assist them in the implementation of science programs. It complements but does not duplicate the resources already available. In this regard, three of Gregory Cajete's books immediately come to mind: *Look to the Mountain, Native Science,* and *Igniting the Sparkle.* Cajete's books are essential to any science program for Native students. *Bridging Cultures: Indigenous and Scientific Ways of Knowing Nature* by Aikenhead and Michell, is a more recent publication whose purpose, as the authors explain and as suggested by the book's title, is "primarily for science teachers and teacher candidates faced with the challenge of teaching the Indigenous knowledge identified in their science curriculum."[22]

Western scientists and educators, by and large, are deficient in their knowledge of Native peoples. They do not recognize the validity of Indigenous knowledge. Any attempt to communicate across cultures, whether to convey Indigenous knowledge and philosophy pertaining to the natural world to Western scientific minds or to explain relevant scientific concepts in terms which are familiar to Native audiences, will require adequate knowledge of both systems and worldviews and the ability to explicate concepts authentically.

With this understanding, I've taken great care to address the question of how some of the concepts of modern physics compare with the metaphysics of many tribes in matters pertaining to the cosmos.[13] By bringing the meanings and terminologies of two systems of knowledge (Indigenous and Western) together as is done here, we can see how they

---

[13] Metaphysics is an area of philosophy that transcends the physical realm. Indigenous metaphysics includes knowledge and philosophy.

correspond to each other and are able to assess what each conveys about the world we live in. In my interpretations I make no attempt to prove Indigenous concepts or to use "competitive" language that invites conflict or controversy. Rather, the aim is to encourage open-mindedness and a hospitable attitude toward new and unconventional ideas that are little understood. Much learning can occur by being respectful when listening to other points of view, by considering what is said, and by engaging in dialogue without arguing.

A word about metaphysics: Metaphysical notions are not verifiable because they are empty of experimental content; however, by transcending the physical realm in search of ways to explain phenomena that are not subject to observation, analysis, or experiment, they are not limited, as in conventional science, to the exclusive reliance on measuring techniques for receiving information. Tribal metaphysics in the context of modern physics acknowledges a live Universe and, therefore, does not display a sterile attitude toward the natural world. The Indian's response to the world involves the whole being through experience, philosophy, participation, interaction, and perception. Some of the terminology of modern physics— networks, relationships, flux, wholeness, connectedness, and so on—is also used in Indigenous contexts.

Natural philosophy, or "Greek science," was the investigation of nature; it was characterized by a *qualitative* mode of inquiry that eventually gave way to a *quantitative* approach, beginning in the 16th century, which evolved eventually into the modern notion of science that emphasizes empiricism, objectivity, and (mathematical) analysis. But another change also occurred and now characterizes modern science: the trend to eliminate metaphysics. More careful investigators noticed that natural philosophers were not always testing their results for correctness before drawing conclusions, even though those conclusions were based on observation. The attempt to rid science of metaphysics has not ceased and may help explain the extreme view, "if you can't measure it, it doesn't exist," which is so prominent today.

Factors to Consider

Changing the paradigm involves other factors as well. First, I believe that the nations of the world need the knowledge that our ancestors amassed through countless centuries but are not aware of it. Native peoples are examples of sustainable science and the nurturing of social systems. It also seems to me that Native students' aptitude and desire for learning about the natural world are partly inherited; they are at least expected to have these traits through a nurturing environment. Physics asks the most basic questions about nature and is consistent with aptitude. The challenge is to educate majority populations and raise their awareness of their dependence on Native peoples and learn to appreciate them.

I do not suggest comparing Western and Native worldviews side by side but allowing each to convey its own truths. Students and educators can make their own assessments and decide what fits into a curriculum.

The second factor to consider is the hegemony of Western science—not hegemony per se but the political threat it represents due to its power to legislate against the welfare of tribal nations residing within nation-states. Hegemony is a threat to Native peoples so long as the majority populace remains ignorant about them. The United States viewed the Indian first as a savage, then as an enemy, a hindrance to progress, and a burden. Treatment of the Indian was motivated by racism or due to ignorance.

One of the aims of this book is to assist in increasing the visibility of Indigenous peoples through awareness of their science, or traditional knowledge, not by "proving" Native beliefs, explaining them in physics terms, offering an alternative theory, or casting Indigenous knowledge into a Western mold, but by providing substantive information to support the affirmation that, prior to recent discoveries, Native peoples had already arrived at the same or similar conclusions about the cosmos without the tools of modern science. This kind of information needs to reach majority populations.

Third are the understandably negative opinions that tribal leaders and other Natives have expressed about science. One problem they cite is the damage to Indian lands as a general consequence of industrialization which, of course, resulted from the application of science and technology. At the core of this problem is the worldview that determines *how* science is applied, resulting in destructive technologies. A related issue is scientific dogma, or *scientism*, stemming from a belief in the superiority of (Eurocentric) science and its empirical methods of validation.

Physics is the basis of all the natural sciences. It is also at the leading edge of discovery at all scales of cosmological research, from the smallest subquantum scale to the largest cosmic scale of stars and galaxies; it is a very difficult field to prepare for as a career path. When I was a student, someone asked "What is physics?" and the reply was "physics is what physicists do after midnight!"

But physics is based on primary concepts that arise directly from sense experiences, or observations, whose study can be personal or impersonal, depending on how the student or investigator adds identity and perspective to the learning endeavor. These include filling the missing spiritual dimension in the study and practice of Western science. No one majoring in a physical science can avoid mathematics. However, for our purpose, which focuses on natural phenomena in an Indigenous context, *many important ideas can be understood by any high school student and integrated with the total learning experience.*

Much of the terminology employed in modern physics lends itself to metaphysical interpretation. A Native philosophy of life reflects the understanding that the Universe is a living, interconnected physical and spiritual network of relationships. If physics concepts and associated phenomena are properly understood and presented, they can help students experience all that the Creator intends about the human being's role in the Creation.

Physics concepts can be understood to a degree without the mathematics in order to show how they correspond to, and accommodate, a tribal worldview. Topics may be selectively incorporated into any school or college curriculum. Several popular books by theoretical physicists are available for the general public but books by Native physicists, designed for Native students, are rare or non-existent and, if they *do* exist, I believe this book still fills a void.

Finally, it seems honorable to do whatever we can within our "professions" to change unwholesome attitudes and ignorance about Native peoples. Because American politicians are forgetful, Indian tribes are always in a precarious position before Congress.

The average American probably does not view the United States as a colonial power. A colonial power gives itself authority to base decisions exclusively on its "rule of law," a legal system that categorically disregards the *moral* and *ethical* aspects of situations which confront it. When deliberating on a social issue, such as immigration, laws are preeminent and usually determine the rightness or wrongness of acts committed; morality and ethics do not apply. A citizenry that is uninformed, whether through willful ignorance or government policy, about the welfare and status of a citizen minority, is aiding the colonial power. Ignorance that begins during youth carries into the adult years and into positions where decisions are made which, although they affect all citizens, are favorable to the ruling majority and unfavorable to the neglected minority.

Allies for Change

It is encouraging to see educators and researchers from the Western tradition begin to accept the more realistic view that Earth is a single interacting system, a trend that may have begun with the advent of environmentalism and environmental science which focuses on human-caused pollution and degradation of the world we inhabit. James Lovelock's Gaia theory,[23] which was first proposed as a hypothesis, affirms that all of Earth's natural systems, both living and non-living, form a single, complex, interactive, and self-regulating system.

Western science is largely unaware of the vital connections that the phrases "all things are related" and "all things are connected" mean to an Indian. These concepts need not be spoken verbatim; as we sit in a circle inside the sweat lodge, they are communicated through the prayers and in our consciousness as we feel connected to the whole cosmos, beginning with our closest relatives and moving outward in each round of prayer to the whole world.

The need for American Indian students and professionals who will interpret these connections, using language that the dominant science establishment understands but *being careful to retain the Indian meaning*, is an important issue. The benefits and conveniences (to humans) that Western science and technology have provided notwithstanding, they are also degrading Earth's condition overall. It is a hard fact that European immigrants developed science *for their benefit* while tribal ecological and social systems have suffered significantly. The U.S.

government's support for tribal issues seems to depend mostly on the political posture of presidential administrations and the U.S. Congress in that conservatives are less friendly.

Among those I consider allies for change are theoretical physicists[14] Albert Einstein and David Bohm. Einstein is, of course, well known world-wide as an active contributor to the early development of modern physics, particularly relativity, until his passing in 1955; his life, philosophy, and work continue to be cited. Contemporary physicists have further refined Einstein's work on relativity and are still referring to his foundational contributions—as well as his mistakes. I personally enjoy reading his explanations of concepts in his own style, albeit translated into English, and they reveal some insights into how he thought.

Bohm, due to his concept of the implicate order and pilot-wave theory, is also much cited in physics literature. Bohm was a colleague of Einstein at Princeton University and also a quantum physicist of renown. Although both are categorized as "Western" scientists, their work strongly corroborates Indigenous knowledge and concepts, including cosmology.[15]

Einstein, Bohm, and later physicists have provided decisive evidences for a Universe that is whole and highly interconnected, utterly overthrowing the notions of mechanism and reductionism, regardless of the extent to which other physicists may still adhere to these obsolete notions. In addition to these evidences, other prominent physicists working on M-

---

[14] Theoretical physics emphasizes concepts, laws, theories, and mathematical models—very difficult work. The goal is to rationalize, explain and predict physical phenomena. All predictions have to be verified; the only way physics can progress.

[15] Cosmology is the study of the Universe in which physics and astrophysics play a leading role. Indigenous peoples also have their own cosmologies (knowledge of the Universe acquired through practice, language, and experiences).

theory and quantum gravity describe the Universe as a "network of relationships" and a "seamless holographic fabric."

Einstein's theory of general relativity[16] treats the gravitational field like a web in which particles of matter are abstractions. Relativity and quantum theory both describe an order of reality that is radically different from the everyday world. Both bolster the concept of unity and wholeness in the cosmos. The idea of the cosmos as a web is, of course, an Indigenous concept. The central theme of Bohm's last book, *The Undivided Universe*, which he completed just before his death in 1992, is wholeness. To my knowledge no Native physicist has used the overwhelming evidences (although Bohm's theory involving the *quantum potential* is not free of problems) to develop a complete cosmology that includes experiential and participatory dimensions and other essential elements of a Native paradigm. Such an endeavor would counteract the fragmented view of the world and might have contributed to an earlier detection of global climate change and other critical issues that our society does not recognize without "scientific" evidence. The warnings of Indigenous elders in different parts of the world have also gone unheeded.

Native authors and educators have identified several concerns about conventional science, including: its inadequacy as a complete and unified system of knowledge; the general attitude and practices of the science community which describe the world in reductionist, unspiritual, and unsocial terms; the arrogance displayed; the implicit and explicit attacks on tribal traditions. The West's perception lacks spiritual guidance and empathy toward non-human forms of life; it also disregards the sacred. These criticisms are valid but will have little influence on the West, in my view, unless convincing solutions are also

---

[16] The general theory of relativity is commonly referred to under the shorter name of "general relativity."

The Condor and the Eagle

proposed. The need for articulating Indigenous perspectives in science within the academic disciplines of the West, has probably never been greater.

Chapter
3

THE ROLE OF PHYSICS

The Most Basic Science

The Greeks of the 4th and 5th centuries BC were mystified about the essential nature and constitution of the physical world. What is it made of? What is the ultimate constituent of matter? The Greek word for nature is *physis*, from which the word "physics" is derived[24] but I hasten to say that "physics" and "nature" had much broader meanings in ancient Greece compared to our modern understanding. One of the reasons for this difference is that the teachings of Empedocles, an important Presocratic philosopher who believed that a correct understanding of nature is connected with the proper way to live, were distorted by later philosophers.[25] Empedocles was also a practicing magician with a strong interest in the practical application of nature and cosmology, healing in particular.[26] Like other Presocratics, he believed that all things are knowingly alive.

Unlike the modern world, ancient Greece categorized some branches of magic as "applied physics," which explains why magicians were referred to as *physikoi* and *physikos* was a synonym for magic.[27] The implication is that, although Aristotle also referred to natural philosophers (*i.e.,* "investigators of nature") as *physikoi*,[28] this term actually had much broader meaning than is commonly thought. Indeed, after reading several sources, it became clear to me that "nature" at that time also had a broader meaning than merely "the material, physical world" and that the early Greeks treated physics as a sub-discipline of natural philosophy—the study of the nature of the world—which embraces *both* natural science and metaphysics. Such was the basic understanding of *original* Greek physics, bearing in mind that the meanings of words change over time and "magic" (*i.e.*, magical powers) is one example.

Modern physics is empirical, not metaphysical; however, the interpretation of phenomena depends on the individual physicist's worldview, not on the practice of physics, which remains authentic and unchanged. Native physicists can interpret the world according to our particular worldviews and even be guided to the phenomena we choose to observe. It has always been true of good science that the interpretation of data varies among individuals while the accuracy of data acquisition depends on the best techniques.

We must not pass over Empedocles' significant reference to nature as the proper way to live; it is about a proper *land ethic*—a philosophy that guides our actions about the use of, and relationship to, the land, and it is just as valid today as it was then. N. Scott Momaday (*The Man Made of Words*) expresses the same basic perspective of a land ethic, saying it is a *moral* understanding as well as a relationship: "We Americans must come again to a moral comprehension of the earth and air. We must live according to the principle of a land ethic. The alternative is that we shall not live at all."[29]

Based on the foregoing, we can safely say that, while physics developed as an academic discipline in the history of Western science, its early Greek beginning was holistic but later followed the tragic path of the West. To Empedocles and other

Presocratics, nature was alive and endowed with spirituality. Their assumptions about nature seem to have been similar to Native peoples' assumptions; they are the basis for my insistence that we should not view physics solely from an academic perspective; the assumption of a live Universe enriches the practice of physics. After all, the study of physics is the study of the nature of the world, the physical/spiritual Universe. The endeavor to discover and explain the forces, energy, movement, matter, and so on, which are at work in the world, becomes a richer experience.

For Natives and non-Natives alike, both educated in the science of the Western world, perceptions of reality and the philosophical interpretation of phenomena will differ; while physics as a discipline of an empirical science need not change, the change that *is* needed will be realized, I believe, only if people with broad-minded attitudes collaborate.

The Role of Mathematics

"Geometry is not concerned with the relation of the ideas involved in it to objects of experience, but only with the logical connection of these ideas among themselves,"[30] explained Einstein. Geometry, in other words, is a pure science because it is unrelated to and independent of sense experience and all subjects outside of geometry. Physics, on the other hand, is not a pure science because mathematical truths are relevant only if they are ultimately related to sense experience; its lowest layer of laws and theories involves actual events that have been observed in space and time. A branch of physics known as theoretical physics focuses on mathematical modeling of theoretical situations rather than specific or real events; its power to predict events and explain phenomena has been substantially demonstrated and numerous examples could be cited. The following are examples: Antimatter was discovered after it was predicted (by Paul Dirac); the bending of light rays due to the presence of mass was confirmed after Einstein's prediction based on the theory of general relativity; black holes, the Big Bang, and other phenomena are also predicted by this theory. Predictions that have not been proved remain open to question, not only because the theory may be flawed but also because predictions are dependent upon the assumptions that have been applied—a matter of extreme importance. A single false assumption can cause the entire theoretical structure or its application to collapse.

Physics concepts are based on measurements that lend themselves to building mathematical models of physical reality. In order to simplify the mathematics, contingencies that exist in nature are commonly neglected in a model but do not affect the model's ability to make realistic predictions. Air resistance and friction, both of which are forces, are two contingencies that are commonly neglected; the nonlinear[17] nature of friction introduces nonlinearity to the equations in the model.

Pure mathematics is a category of abstract mathematics that is independent of experience; however, it plays an important role. For example, without Riemannian geometry (a topic in advanced mathematics that deals with curved surfaces in multiple dimensions), Einstein would not have been able to complete his general theory of relativity.

Physics as an applied science endeavors to unify our knowledge of the laws of nature, which we assume are already unified. Mathematics expresses the laws of physics but it is not the primary reality. Eminent physicists like Einstein, Bohm, and others have explicitly stated that all physics laws are tentative.

---

[17] A nonlinear system (introduced in chapter 3) cannot be described by equations of first degree, like $F = kx$, where the variable x is of first degree and the equation is linear; x^2 would make it of second degree. Nonlinear systems do not have a constant of proportionality.

The question whether the world is fundamentally mathematical was entertained by the early Greeks.[31] Some, like the Pythagoreans, believed that geometry is inherent in nature; that God created the universe according to a geometric plan. Mathematics is perceived by some as a lens through which we see the real world, but I view mathematics as a tool. Nor can it be claimed that mathematics is Western; for tribal peoples like the Maya and others, mathematics is part of their cosmologies. What *is* Western is how mathematics was developed and how it is applied. Numbers have a different significance in Indigenous cultures. According to Mayan holy man Hunbatz Men, ancient Mayan religion was scientific and mathematical, arising from their sense of human ethics.[32]

Primary Concepts

Observations are also referred to as sense experiences because they are witnessed directly, detected by instruments, or perceived by our senses; they form the first layer in the logical structure of physics. Einstein explained that a complexity of experiences gives rise to primary concepts, which in turn lead to propositions and laws that are expressed mathematically and verified (or falsified) through measurement. All other notions have meaning only as they are connected by theorems to the primary concepts.[33] Physics is connected to the real world. Since sense experiences are subjective, it can be said that science, or knowledge, begins there. Knowledge can arise from a single experience but one does not build a science from a single observation.

"Theory" (from *theoria*, or theatre) often means "insight" in common usage. It may also refer to the rigorous explanation of a concept or a principle, such as the theory of relativity, the theory of networks, or the idea or proposition itself, such as when someone theorizes or asks, "what is your theory about this?" The statement, "That's only a theory," usually expresses doubt

The Condor and the Eagle

about an idea. Observed phenomena are sense experiences because they pass through human sensory channels (or a substitute, such as an unbiased instrument). Facts are related to theories in the following manner: "The sense-experiences are the given subject-matter. But the theory that shall interpret them is man-made."[34]

Physics as an Ally

Of all the natural sciences, physics asks the most basic questions about the forces, energy, movement, and matter that dominate the natural world. Some questions are purely academic from the perspective of science students who are motivated mainly by curiosity; in the majority society, science is for scientists and is highly specialized but in Native communities knowledge about the natural world has been, historically, a way of life that is often related to survival; it is practiced by the community living *with* the land, not just *on* it. Local knowledge is profound and depends on complete familiarity with the landscape.

Scientists from the Western tradition can be great allies in bringing the natural sciences in contact with traditional knowledge so as to change or replace the current paradigm— and many are. As the most basic science, physics connects sense experiences with theory and forms them into laws that can be tested. The fundamental units of measurement—time, space, mass, temperature—from which every other unit in science, not only physics, is derived, are introduced early in an elementary physics course. Other branches of science use measurements to capture physical data that are expressed in units derived from physics because they are ultimately a form of physics. For example, chemistry is actually the physics of three particles (proton, neutron and electron) which are influenced by the electromagnetic force. The unit of force is the Newton, which is derived from Newton's second law, $F = ma$. If m is measured in kilograms (kg) and acceleration, a, is measured in meters per second per second (meters per second squared, or m/s^2), then force equals kilograms times meters per second squared (kg-m/s^2), which is expressed in its fundamental units: mass, length, and time. The attractive force between a positively-charged proton and a negatively-charged electron, and the repulsive force between two like-charged particles, are examples of the electromagnetic force.

Relativity theory and quantum theory describe cosmic domains that are vastly different from the classical realm. Bohm was probably the only physicist to propose an alternative approach based on the concept of "undivided wholeness" but not the only one by any means who recognized the evidence of wholeness. His work, beliefs, and concepts, and those of Einstein and others, corroborate several tribal principles and beliefs.

Many physics concepts can be understood without mathematics but they cannot be ascertained or described without mathematics. When the concepts are expressed mathematically, they form complete concepts in the physicist's mind and complete sentences in the physicist's language. The mathematics of physics may seem unattractive; nonetheless, much of the science it represents is experienced in our daily lives in the form of forces, energy, matter, light, gravity, and so on.

We base of our lives largely on conventional ideas, unaware of the hidden realities that have been verified by science and experienced in Native traditions. A practical question is whether we want to live an *illusory* existence that ignores these realities or to be consciously connected to, and aware of, the mysteries of the cosmos. As individuals and as nations, if we have the ability to perceive the differences but habitually ignore them, can we claim to have a realistic view of the world, or are we basing our lives on beliefs and ideals?

The approach I have taken in this book is to describe modern physics and its historical development from an American Indian perspective and, if relevant, to point out the implications. My own experience and knowledge of Indian philosophy is limited, but this is true of all Natives, for each tribe has its own traditions. Despite the variances among tribes, there are also shared beliefs and experiences. My hope is that more experienced and knowledgeable Indigenous elders and others will use what I present here to make deeper connections that support efforts to create a global consciousness and achieve a global conscience which will move into action.

Native educators need not distance themselves from reputable Western scientists while pointing out the latter's prejudices. Einstein and Bohm were both reputable physicists whose disagreements with others were not due to arrogance or conformity. Einstein was proved wrong twice on quantum theory; he openly admitted one mistake and passed to the spirit world before he was proved wrong on the second one, but he left a challenge that kept physicists busy for a long time.[18]

---

[18] These issues involved Heisenberg's Uncertainty Principle and the phenomenon of non-locality, which Einstein referred to as "spooky" action at a distance. Einstein and two of his colleagues presented a challenge to Neils Bohr and others that came to be known as the EPR paradox; the challenge failed, quantum physics was validated, and Einstein was proved wrong.

Chapter
4

THE WORLD WE LIVE IN

A view of the world from a classical perspective will help us more clearly see the dramatic difference that modern physics makes to our understanding of the real world.

American films that portray historical events often show scenes of the Old West or crime in the cities. Science fiction films about the future, on the other hand, are more imaginative because the events have not yet occurred. Many of them show scenes that require a great deal of technology and few natural landscapes such as mountains, trees, and rivers. Do science fiction authors depict what they realistically expect the future will be like?

Many futuristic scenes depict adventures into space or a completely automated world that requires sophisticated and obviously expensive gadgetry. In some films, the only visible skin on the body is the face and possibly the neck, as well as features that are probably meant to portray highly evolved humans. In this imagined world, possibly on another planet, there are no natural landscapes. I would not want to live in such a world.

An Illusory World

It is assumed in this kind of future that unlimited resources will always be available and that Earth will sustain exorbitant lifestyles indefinitely, which of course is an illusion and, after all, the film was produced for entertainment—and for profit.

But let's look at reality: the problem of humanity is to maintain an absurd self-centered posture and enjoy a make-believe world of human ingenuity and greed while millions suffer and Earth's life-sustaining systems deteriorate.

Meanwhile, though our society is informed in many ways, it is so accustomed to the visible and audible world that few of us are even vaguely aware that the visible world is fully sustained by an invisible micro-world. We deceive ourselves if we think that what is visible is real and what is invisible is not real or important.

Actually, the converse is true; *everything we see, which comes through our senses, depends entirely on the real world that we don't see*, but for me as an Indian, this kind of faith defines reality. It is a faith that says "I know" when conventional thinking says "I believe." In the Bible we find an equivalent statement: "Now faith is being sure of what we hope for and certain of what we do not see. This is what the ancients were commended for. By faith we understand that the universe was formed at God's command, so that what is seen was not made out of what was visible."[35]

Human experiences involve interactions between the spiritual and physical realms and between what is invisible and what is visible. Some tribal people say that the spiritual realm is *primary*; it is invisible and holds deep mysteries and the latter is secondary, since things that are visible were first invisible and unaltered by the physical. Of course, many physical things are not visible to the unaided eye but they are physical, nonetheless. Some people believe that these two realms correspond to Plato's theory of *form* and *matter* (discussed in a later chapter) but these concepts were not derived from Plato.

Tribal creation stories commonly include animals and people who transform themselves into other forms; it's called "shape shifting." Vine Deloria, Jr.'s last work, *The World We Used to Live In*, is a collection of eyewitness accounts that illustrate the immense power of medicine men during a former pre-contact era. "It is fair to say that the overwhelming majority of Indian people today have little understanding or remembrance of the powers once possessed by the spiritual leaders of their communities."[36]

Becoming increasingly drawn to the noise and artificial gadgetry of the modern world implies increased dependence on automation, less dependence on human skills and potential, and less aware of spiritual realities. I am convinced that civilization has so alienated us from the natural world that we have lost many of the abilities and spiritual powers that past generations utilized and enjoyed. We are out of practice, no longer capable of performing certain tasks which our human relatives who still live interdependently with nature can still perform, such as navigating the oceans and predicting weather without instrumentation. Neither they nor our non-human relatives have lost their natural powers. Animals are more sensitive than humans even to events and sounds that do not depend on hearing or smelling.

It is also common in our everyday world to witness events that would be illusory if we didn't know better. We say the sun rises and sets but we know it's the earth that rotates on its axis instead of the sun circling the earth. We imagine a highway as flat as a map lying on a table or posted on a wall but we know that the road, if extended indefinitely, would circumscribe the round earth. We enjoy the serenity of a forest and the restful silence of our living room and disregard the earth's movement through space at incredible speeds and in different directions simultaneously.

Everything is in motion—the earth relative to the sun, the sun and solar system relative to the galaxy, and even our galaxy relative to other stars and the billions of other galaxies—while here on the earth we pay little attention to what the cosmos is doing. Light has unusual qualities; we ignore the time it takes for it to propagate. The beam from our flashlight at night or the flash from a lightning strike seems to arrive instantly yet we know it travels at a finite speed, though much faster than other speeds, including the speed of sound. We know that the starlight in the night sky began its journey long before we perceived it but we still say "they shine" as if it were now.

Events like those above are easy to visualize and explain with the common-sense laws of classical physics. The motions of planets in our solar system and all celestial bodies are subject to external gravitational forces that are governed by laws of motion which are approximated to a high degree by the equations of *classical mechanics*—Newton's equations of motion. Using a different approach that takes the special theory of relativity into account, Einstein derived the equations of general relativity theory that give us a deeper understanding of gravitation. Newton's laws are still practical in cases that do not involve large masses, speeds close to that of light, or the quantum scale. In the classical world, if the state of a physical system (given in terms of position, momentum, and initial conditions) is known at any instant, its state at another instant can in principle[19] be precisely determined.

Newton's laws are peculiar in that they assume that objects (typically referred to as "bodies" in physics) are represented by

---

[19] Newton's laws are not affected by the complexity of the physical system but the equations become more difficult to formulate and solve as the number of objects increases.

hypothetical points in space with mass but no dimensions. In real situations, bodies are three dimensional; they are treated mathematically by assuming that the forces are applied at a unique point on the body called the *center of mass*, where all of the mass is considered to be concentrated.[20] All measurements, such as mass, time and distance, are approximated but adequate for predicting orbits and sending rockets and cargo on successful space missions. In situations where Newton's laws are not applied, such as manufacturing parts for an engine or a wrist watch, measurements must of course be precise.

The laws of physics are attempts to represent and understand physical reality. However, unexplainable phenomena can also be shown to exist and this is where our discussion must now take a major turn. The advent of the quantum and its strange behavior in 1900, followed by other major discoveries during the two decades that followed, revolutionized physics and raised new philosophical questions about the nature of reality, matter, energy, time and space. At submicroscopic scales, knowledge of the world cannot be obtained through direct observation, nor can we visualize nature's behavior. The realm of modern physics, also known as the "new" physics, treats events that occur at small scales which must be modeled mathematically due to their unvisualizable and counterintuitive nature.

In the chapters that follow I provide details and rationale from modern physics and apply my understanding to explain how I believe they relate to Indigenous ways of knowing. Suffice it to say for now that we live in a macro world of illusion, largely disregarding the reality that lies beneath the things that keep us occupied. Most of the new and important

---

[20] The center of mass is not always a point *on* the body; it could be in space, as in the case of a square picture frame.

The Condor and the Eagle

concepts of modern physics will be introduced in chapters 6 and 7.

Why Humans Explore

The question of what motivates the human quest for knowledge would yield a plethora of varied responses if we seek answers based on each individual's preferences but they would not be relevant to our purpose. We are interested in what drives people to explore the unknown, what they expect or want to learn about the world, and what they do when they gain that knowledge. The answers depend to a large extent on values that were forged over time by a people with a common history, who shared the same land, language, and experiences for generations and built a culture. Cultural values are deeply seated.

Whereas Western culture has followed a historical path that can be traced to the Greeks, each tribe has its own story of origin independent of Western culture. Needless to say, Western and non-Western cultures differ widely. Even the Indigenous peoples of the Americas differ among themselves on worldview, language, and way of life, largely due to living in diverse local environments. But their views of the natural world also have much in common.

The following example may help to illustrate American culture's deep motives for exploration. It is perhaps during times of crisis that a society's most deeply rooted beliefs and passions are expressed. In the wake of the Columbia shuttle disaster of February 2003, which claimed the lives of seven astronauts, National Aeronautical and Space Administration (NASA) officials and other scientists, science educators, news editors, and columnists were already thinking beyond the grieving period and expressing the firm conviction that the space program should continue. They seemed to feel so threatened and troubled by the possibility that the space program might be terminated prematurely that they did not wait in reverent silence for fellow Americans to get over the initial shock from the tragic event. Instead, they deemed it important to advocate *immediately* for the continuance of the space program.

As to the reasons for continuing the space program, articles repeated themes that reflected typical Western values: a) to follow the natural instinct of human nature (or the human spirit or the human heart); b) to push the envelope of science and soar to new heights because c) the unknown and the unexplored world is there. Some spokespersons added the phrase, "though not always hospitable," implying that risks are always present and accidents like this one are always possible, as if to say "it's still worth it"; continuance is a foregone conclusion.

Two weeks shy of the first anniversary of the shuttle disaster, the White House announced President George W. Bush's plans to return to the moon and eventually to Mars. "Mankind is drawn to the heavens for the same reason we were once drawn into unknown lands and across the open sea," he said. The announcement also drew a parallel between Lewis and Clark's expedition into "new" lands and America's venture into space, both done "in the spirit of discovery" (*i.e.,* because it's there).

Dominant notions about human nature and the goals of science reflected in the above statements stem from assumptions and doctrines that are deeply seated in American culture. Some of the claims, such as the assumed parallel about the Lewis and Clark expedition, are erroneous because the explorers came upon a world that was already inhabited by fully functioning societies; those territories were not unexplored or unknown. The tribes were intimately familiar with the landscapes that had nurtured their people long before the arrival of Europeans. The land was "uncharted" and "wild" to these foreigners because they thought of nature as strange, wild, and hostile. Two hundred years later, Americans still have not connected with the land in a loving way.

The triumphal language is reminiscent of the doctrine of Manifest Destiny. In the Western worldview, humans are seen as progressively overcoming barriers along a path that will achieve the ultimate human potential. But notions of conquest, linear progression, and human hegemony over nature do not reflect sustainable science. Americans express these views with great confidence, but from an Indigenous perspective, they are not valid.

In contrast, Native peoples focus on survival, the welfare of future generations, the continuous existence as a people, and related issues. By and large, they have not concentrated on industrialization, empire building, exploration for its own sake, or large-scale activities that are inconsistent with the purpose of continuing to live in a proper relationship with the natural world.

In summary, Western civilization and Indigenous peoples forged separate and distinct historical paths and for vastly different purposes; the consequences are not a question of who suffered more for doing right but who will be judged for doing wrong, and only a righteous Judge can settle it. More than two millennia ago, when the Greeks and Indigenous peoples seemed to share similar concepts about nature and about the unity of knowledge, the two paths may not have been far apart, particularly since all peoples originated as tribal societies.

Perception

The U.S. government's lack of interest in tribal issues helps to explain how little the citizenry knows about the tribes—their socioeconomic status, causal history, and knowledge of the natural world—so it's not surprising why so many Americans also display negative attitudes. The low priority given to vital tribal issues, opposition to those issues, the rare mention of them on the nation's agenda, and the lack of media coverage by corporate-owned networks all play a role in creating suspicion and turning the American Indian into an enigma.

Indians have learned to be vigilant, in view of the questionable federal policies, national attitudes and assumptions, and disappointments. The often-expressed phrase, "just another broken promise," indicates the betrayal that Natives sense when the government fails to deliver on its obligations. Frustrations are often experienced when we a Native news source reports that certain issues remain unresolved or that negotiations have deteriorated.

University campuses, where I spent a total of 44 years, are places where people from different cultures and backgrounds have the opportunity to come together, learn about each other's values and backgrounds, and establish lasting relationships. My wife and I met families from several countries; our children played with theirs, we were invited into their homes and to their country, and our lives were enriched.

Americans, on the other hand, seem to be less amiable toward Indians. There is the perception of the Indian as a "conquered race," which is further encouraged by old Hollywood films; other perceptions abound: that Indians are holding onto an obsolete culture, are unqualified to own land, are unable to govern themselves, and are useless to society. Christian talk shows reveal additional misconceptions: that Indians don't deserve government help, since they were "defeated"; they live on welfare and can get a free education; they know how to "work the system" to get what they want; Indian reservations should be eliminated; Indian treaties are out of date; they are a burden to society; and so on.

The U.S. government once viewed the tribes as enemies and obstacles to progress. America was in fact built on the denial of all that the Indian represents: values, language, land, rights, identity, culture, spirituality.

The concept of diversity appeared on my college campus in the 1990s and the slogan, "respect the differences and celebrate the similarities," with reference to the fair treatment of all ethnic groups, soon followed. Unfortunately, it disregarded the comparative histories of immigrant and native populations and essentially conveyed the message that everyone should learn to get along with each other; it focused on "we are all here now" and ignored the vital question, "*how* did we all get here?" It also categorized American Indians as ethnic "minorities," which hid our official relationship to the U.S. government through Title 25 of the U.S. Code and, very significantly, it gave the white majority the opportunity to redefine *diversity* in ways that accommodated their own views while avoiding the issue of racism. Colleagues in my department and high-level management could not hide their resentment toward the university's emphasis on diversity. Some of them approached me occasionally to challenge the comments I had made in meetings on certain issues. "*Everyone* is diverse," they would say, pointing to how people are different, including their tastes, to clarify how they defined "diversity." The work environment was antagonistic and indisposed to learning from the values and strengths of other people.

Why is it this way? The absence of a historical perspective has created a wide gap in the minds of citizens who cannot understand why certain ethnic groups in the United States experienced more abuse than others or why some are still underrepresented. Because corporate news networks have their own agenda, it is necessary to seek information on various issues from independent sources, and since this takes extra effort, many are simply uninformed on those issues. Howard Zinn, the late American historian, educator, and author, explained that the U.S. Constitution was drafted by a wealthy elite who owned land and had slaves and secured their own future, as well as the future of their class of Americans; they excluded certain groups from the constitution's protection: black slaves, poor whites, women, and of course American Indians. As is well known, constitutional amendments became necessary for groups that were previously excluded.

Immigrant Americans from Western Europe generally don't view themselves as colonizers. They seem to believe that the United States broke no laws when forming the nation. It is understandable for Americans to be proud of a heritage linked to a country which was once their home or the home of their beloved ancestors. The younger generation have a shorter memory that may be limited to recent childhood experiences. But perhaps their sense of pride swells as they learn about their ancestral country or they listen to parents and grandparents tell stories of what it was like to become separated from home, relatives, and friends, as well as the dangers they faced, the sorrow they felt, and perhaps how well they were welcomed to another country that they now call home.

Older generations of American immigrants can best imagine the experiences of tribal peoples who were forcibly left their homelands and marched to places that were unfamiliar, where they could no longer live off the land, their stories and memories no longer connected with the land, and their language no longer related to the land.

Perspective

American Indians and many other Natives around the world share experiences related to colonization and colonialism, including the transfer of land title to Europeans. Becoming dispossessed of land does not imply that Indians have forfeited that loss in mind and spirit. Many still think of it as Indian land, given by the Creator who has never decided to give it to others, and he is the final arbiter in the matter. I have never heard of a tribe complaining that it doesn't like its homeland or that it is an inadequate place for the tribe to live. I do not accept the notion that the tribes were destined to be displaced and humiliated or that conquest is simply a part of the human history ("all nations conquer"). This kind of thinking disregards the Creator's intent and original instructions to the nations he created and nurtured.[21]

The perspective I try to maintain is based on the constancy of things. I'm not a conformist; what happens outwardly and historically has not changed certain aspects about our Indian identity or our original purpose which was determined by a higher order. The preemption of traditional tribal governments by nation-state governments that have no direct relationship to the land has created conflicting situations for tribes and individual Indians who once enjoyed complete freedom to follow their traditional ways. Examples include the possession of eagle feathers and other ceremonial objects, national border issues (such as crossing ceremonial objects to or from Canada and Mexico or simply crossing the border), reduction of ceremonial rights and privileges for Native inmates, and various jurisdictional issues.

When Primer Minister Keating of Australia (mentioned in chapter 1) spoke at that country's 1993 celebration of the International Year of the World's Indigenous People, he gave consideration and honored their viewpoint as a colonized people and asked his audience to imagine what it's like to be in the place of the Aboriginal people. Rather than make a single

---

[21] This issue is discussed in more detail in chapter 9: "Land and Spirituality."

general statement, he identified each abuse individually, demonstrating his insight and apparent familiarity with that country's Aboriginal history. What he described was similar to the experience of American Indian tribes. I listened to his entire speech on video and copied the following captions:

> It would help if we non-Aboriginal Australians imagined ourselves dispossessed of the land we lived on for 50,000 years, and then imagined ourselves told that it had never been ours. Imagine if ours was the oldest culture in the world and we were told it was worthless. Imagine if we had resisted this settlement, suffered and died in the defense of our land, and then told in the history books that we had given it up without a fight. Imagine if non-Aboriginal Australians had served their country in peace and war and were then ignored in the history books. Imagine if our feats on the sporting field had inspired admiration and patriotism and yet did nothing to diminish prejudice. Imagine if our spiritual life was denied and ridiculed. Imagine if we had suffered the injustice and then were blamed for it.[22]

I rarely use the term "Native American" but it's only a matter of preference. It seems to place the conversation in a European framework, since it originated from that perspective, and it gives preference to a U.S. American identity over a Native identity; a more correct term might be "American Native" (Native of the Americas); however, this is not a suggestion. "Native American" also sounds hyphenated, as the following terms once were: Mexican-American, African-American, Chinese-American; these are immigrant designations similar to

---

[22]Paul J. Keating, prime minister of Australia, Redfern Park, December 10, 1993, at the launch of Australia's celebration of the International Year of the World's Indigenous Peoples. The quoted text is from the speech at URL = < http://www.youtube.com/watch?v=x1S4FleuzTw>. The transcript of the speech (at this writing) is at URL = <http://australianpolitics.com/1992/12/10/paul-keatings-redfern-speech.html>.

"European-American." Natives typically don't refer to themselves as Native American; they call themselves by their tribal name.

I once viewed myself as a victim until I realized it was not a healthy attitude to maintain, although the experience I went through may have been necessary in order to arrive at a mature state psychologically, one that strengthens rather than weakens. Contrary to what I've heard, I do not find much bitterness in Indian country; tribal communities have, in fact, been a source of healing for me and for my wife.

With a proper perspective in the face of incontrovertible historical facts we can become strong advocates of what is good instead of spiteful and malevolent victims. My own experience was lengthy and traumatic; every issue had to be dealt with but not all at once. Each tribal citizen's experience is different; since I was not raised in my tribal culture, my response to the outside world was different from that of traditional Indians. My training in Western science undoubtedly would have been vastly different had I known Indian ways all of my life. This is not a lament and I have thankfully learned to function in multiple cultural worlds.

The Vast Universe

I began my senior year in college as a physics major in 1958 perplexed by basic questions. Is there a physical boundary marking the edge of the Universe? If so, what is on the other side? Why are we (humans) here, considering that we are so small and our time on Earth is so brief? The "edge" is moving with the expanding Universe. As to what is on the other side, it depends on whether anything exists beyond space-time. So, as far as we know, it is still a mystery. Has the Universe always been expanding? Is it unbounded, finite, or spherical? How did it come into existence?

The Big Bang theory[23] is not free of problems. The Universe did not necessarily begin with the Big Bang because this event could not happen without pre-existent laws, and energy, and primordial conditions.

The shape of the Universe (*i.e.*, the curvature of space-time) is determined by the ratio of its density, defined as the amount of matter (mass) and energy per unit of volume, to a quantity called the critical density—the value at which the Universe would neither expand nor contract. If the ratio, called the density parameter, is equal to 1, we have a flat Universe; if greater than 1, the Universe is spherical; if less than 1, it is hyperbolic (saddle-shaped). It is interesting that the actual density of the universe is observed to be very close to the critical value; *i.e.*, our Universe is apparently flat. The question of how the initial density came to be so closely fine-tuned to this "special" value is known as the "flatness problem."

A further complication is that the total mass in the Universe is comprised mostly of dark matter, while dark energy dominates the total energy. The Universe's fate thus depends on its density; if it is greater than the critical density, it will eventually collapse, an event called the "Big Crunch."

Einstein believed the Universe was static (neither expanding nor contracting but in a "steady state") and eternally the same, but he was wrong.

In 1962 I was teaching physics and working toward a master's degree. During my master's oral examination, the people on my committee, who were also colleagues, played a practical joke on me by asking questions that were unsolved problems in physics; as I struggled to answer, they were enjoying it.

---

[23] The term "big bang" comes from Fred Hoyle, who advocated the steady-state model of the Universe.

The Condor and the Eagle

As a student, I thought of the vastness of the Universe with a sense of awe, yet I was occupied with the world within my reach. In my lifetime, I have not met other scientists like Albert Einstein, who was moved emotionally by the complexity and harmony in the Universe. "My religiosity consists of a humble admiration of the infinitely superior spirit that reveals itself in the little that we can comprehend about the knowable world."[37] His was a deeply emotional conviction of the presence of "a superior reasoning power, which is revealed in the incomprehensible Universe" and "a spirit vastly superior to that of man [which] is manifest in the laws of the Universe."[38] His personal life, as revealed by his grandson Bernard[39] and his biographer Walter Isaacson (*Einstein: His Life and Universe*), is also worth noting.

Our planet is of course very small compared to our sun. The sun is large enough to hold 1.3 million earths inside it. Yet the sun is not large compared to other stars. If we could stand on the largest known star in the Universe, the sun would look like a speck of dust or a grain of sand. The largest and brightest known star, VY Canis Majoris, is three billion kilometers in diameter, more than 2000 times the diameter of the sun. More than 200 billion stars populate our Milky Way galaxy—a swirling, disk-shaped cloud—and there are hundreds of billions of galaxies. It estimated that there are over one trillion stars in the Andromeda galaxy and a vast number of galaxies.

Looking in the opposite direction in terms of scale, the nucleus of an atom is 10,000 times smaller than the atom itself, like looking at an apple in the middle of a football field. An atom is 100,000 times smaller than anything we can see without a magnifier such as a microscope. Everything too small to be visible, from atoms on down, is the domain of quantum physics, which deals with phenomena (particles) at microscopic and sub-microscopic scales, and is also called *microphysics*. Newton's laws of motion are not valid at this scale. There are even smaller particles at the scale of subquantum physics.

Deep down at the root of nature—below the level of matter, wherever that may be—is the Planck scale, on the order of 10^{-35} meters—so tiny that there's no hope of probing this length of scale directly in the foreseeable future. To get an idea of how small the Planck scale is, let's compare it to the size of a proton, which is already very small, since it forms the nucleus of an atom. The size of a proton is *larger* than the Planck length by roughly the same factor that the state of Rhode Island is larger than the proton! Lest anyone believe that the quantum world[24] is too small to have any noticeable effect, I must emphasize that the entire Universe, despite its vastness, is sustained by the smallest parts. The functioning of the Universe depends on actions that occur—and are always occurring because they never cease—at the quantum level.

Earth, the world beyond Earth, and the world of ordinary daily experience consisting of objects of all sizes that can be touched and seen with the unaided eye—airplanes, houses, plants, animals, humans, and so on—are the domain of *classical physics*. All objects in this domain can be described and predicted adequately by Newton's laws of motion and equilibrium, except that objects traveling close to the speed of light are subject to the laws of relativity. When we speak of an "object" with reference to Newton's laws, we mean something that is three-dimensional and is occupying space at a particular instant in time.

A point I want to make about these spatial domains is that for countless generations, Native peoples have known, and more recently have been reiterating, the important principle that

---

[24] "Quantum world," which appears frequently in this book, refers to the minuscule scales at which quantum actions occur. At this scale, the laws of classical physics do not hold.

all things in the vast cosmos are interconnected and interdependent. Modern physics has corroborated this and other principles that Native peoples had already known and trusted, such as the notion that events that the notion that "all things are connected" and the principle that the visible world is sustained by an invisible world, which the Haudenosaunee (the Iroquois Six Nations) express in terms of the species of grass arising from the spirit of the grass: "A blade of grass," they say, "is an energy form manifested in matter—grass matter. The spirit of the grass is that unseen force which produces the species of grass, and it is manifest to us in the form of real grass. The Creation is a true, material phenomenon, and the Creation manifests itself to us through reality. The spiritual Universe, then, is manifest to Man as the Creation, the Creation which supports life."[40]

These are not exaggerations or analogies; they are physical principles. But why should it matter? Because, sadly, here on Earth, where Creator chose to give birth to and nurture his nations, the gifts that he bestowed on humanity are being used to exploit and to destroy. Yet we see our non-human relatives using their gifts to harmonize with Creator's laws. I once heard a story by a Nez Perce elder about two salmon that were swimming together up the Snake River on the way to separate spawning grounds. They eventually reached the confluence of the Snake and Clearwater Rivers, where they were destined to separate. "Did they stop to decide whether or not they should continue to swim together?" the elder asked. "No," he said, "they followed Creator's laws and went their separate ways to fulfill their respective destinies." But we humans often use our freedom of choice to violate these laws.

Chapter
5

ANCIENT KNOWLEDGE

T he West prides itself in its accomplishments. One subject of interest is the accelerating growth rate of human knowledge and its causal factors, with technology at the leading edge. It is said that knowledge accelerates because each generation benefits from the cumulative knowledge of the previous generation.

However, the ancients do not seem to have followed this pattern of accumulating knowledge, considering that they lived in diverse local environments. In any case, the immense knowledge that they attained and was later disseminated to the whole world is worthy of acknowledgment. What they were able to accomplish is truly amazing and, in my opinion, should invoke our humility.

Megalithic Structures

Paul LaViolette (*Genesis of the Cosmos*) tells us that the earliest written record of Egyptian mythology describing the creation of the cosmos comes from the Pyramid Texts; these are inscriptions which date from circa the 25th to the 22nd centuries BC. The myths[25] they describe are believed to originate from a much earlier time, from the pre-historic era preceding the rise of Egyptian civilization.[41] He also asserts that some of the large megalithic structures built by ancient peoples are so immense that their construction must have utilized technologies far in advance of those available to known ancient civilizations;[42] as examples he cites the pre-Inkan fortresses, temples, and walls found at Ollantaytambo, Tiahuanaco, and Sacsayhuaman in the Peruvian Andes. At Ollantaytambo, which my wife and I visited in 2011 when we participated in the World Indigenous Peoples Conference on Education (WIPCE) held in Cusco, Peru, stones weighing between 150 and 200 tons were lifted by unknown means to the top of a 1500-foot cliff and exactly fitted together. At Sacsayhuaman, where WIPCE's opening ceremony took place, there is an immense theatre built entirely of stones weighing 100 tons and perfectly assembled. This was the fortress of Sacsayhuamán, says LaViolette, where the Inkas performed sacred rites and held festivals. This fortress also served as a refuge in cases of emergency, thanks to the imposing walls that protect it.

Like many others who are amazed at the incredible intelligence and engineering skills required to perform such feats—skills which are now lost—LaViolette asks whether "these [could be] relics of a technically advanced race that once inhabited the globe and met its decline during a period of climatic turmoil." During our visit to these two places near Cusco, our bilingual guide explained how the ancients did it: They cut the stones in a quarry below and then transported them up the mountain. He, like other Peruvians, politely expressed annoyance at the suggestion that ancient (Indigenous) Peruvians were incapable of accomplishing such feats. Many others from the North, like me, have heard similar statements

---

[25] Stories and legends

expressed in documentaries that are cablecast on the History Channel. While we in the 21st century are not able to explain the phenomenon, we must ask why modern scholars are so willing to accept the possibility that Earth was visited by smart extraterrestrials, or ancient "cosmonauts," and so ready to dismiss the possibility that ancient Natives possessed the necessary scientific and engineering knowledge. If the latter is true, the disappearance of that knowledge is so unfortunate for our generation because of what we might have been able to use wisely.

In the early 16th century, soon after the arrival of Columbus, the Spaniards, led by Hernán Cortés, more or less repeated this same sad story of destruction when they conquered the Aztec Empire and reduced to ashes all of the sacred Aztec and Mayan writings they could find, except for some fragments of the books of *Chilam Balam*[26] and others that were recovered, representing an unknown amount of knowledge that was lost forever. It's worth asking, again, how our generation might have benefited from what they knew and taught.

Non-Indians, conquistadores and colonists alike, have the bad habit of believing they are superior to Indians. When the first Europeans arrived, they spoke highly of Native hospitality, good manners, and skills; a few years later, the Natives were suddenly "savages," a label that survived two more centuries until the signing of the Declaration of Independence, where we find an even more pejorative expression: "merciless Indian savages."

---

[26] Sacred books of the Maya of Yucatan, written in the Yucatan-Maya language and named after their last and greatest prophet. *Balam* means jaguar and *chilam* means priest.

Ancient Libraries

Deloria (*Evolution, Creationism, and other Modern Myths*) cites vast collections of ancient literature that were destroyed in wars and religious purges. "[I]t is doubtful that we possess even a thousandth of the knowledge that was gathered together, analyzed and edited, subjected to additional commentaries, and regarded as reliable by ancient peoples. ... These books were not by any means the novels, 'how to' guides, and children's books that make up the bulk of our modern libraries' holdings. They must have represented hundreds of thousands of years of careful thought..."[43]

In order to give some evidence of the knowledge of ancient peoples, Deloria gives estimates of the number of books that were once stored in various libraries and were lost. The total number of books at Pergamus (in Asia Minor), Carthage in 146 BC, the Druid libraries in the 2nd century BC (thousands of scrolls on philosophy, medicine, astronomy, and mathematics), Byzanthium in the 8th century AD, and Alexandria, is approximately 1.7 million, not counting the thousands of scrolls in Druidic libraries. Besides the books and scrolls, there was a Sumerian tablet listing constellations that went back to circa 11,000 BC and star records from the Maya, Aztecs, Olmecs, Inkas, Egyptians, Greeks, Chaldeans that went back 370,000 years [*sic*]. Deloria writes: "These numbers are startling and raise questions that we should have faced honestly when this data was discovered."[44]

Inter Press Service reported in February 2005 that the Iraq National Museum, which contains precious relics from the Mesopotamian civilization, suffered losses from damage and looting during the U.S.-led invasion in 2003 and subsequent occupation of Iraq that included one million books, 10 million documents, and 14,000 archaeological artifacts. It was the biggest cultural disaster since the descendants of Genghis Khan destroyed Baghdad in 1258, according to what Venezuelan writer Fernando Báez told IPS.[45]

Pre-contact Science

Fifteen thousand years ago, Indigenous peoples from the Arctic Circle to the tip of South America began to introduce inventions enjoyed by people throughout the world. They made contributions in agriculture, architecture, astronomy, medicine, ecology, engineering, aquaculture, horticulture, and other fields. These are described in The *Encyclopedia of American Indian Contributions to the World*[46] which includes more than 450 inventions and innovations. This publication is of the highest quality in every respect; the first time I opened its pages, a sense of awe and pride came over me. I thought of how little the world knows about their enormous knowledge and intellectual capacity and ingenuity. To qualify for inclusion in the encyclopedia, each item or concept it must have originated in the Americas, must have been used by the Indigenous people, and must have been adopted in some way by other cultures. Until the late 19th century, only anthropologists and archaeologists knew about these inventions and their origin, yet even today Natives are not clambering for recognition.

My main purpose here is to introduce readers to this valuable resource and to provide some examples from it.

- Electricity (c. 200 BC – AD 600): The Moche people (Peru) invented electrochemical production of electricity for electroplating purposes. They formed an acid solution using water and chemicals, which naturally occurred in the environment, then dipped copper into it to produce electricity.

- Medicine: sophisticated healers relied on botanical drugs that remain in use today, including GUAIACUM, IPECAC, KAOLIN, QUININE. They also used antibiotics and antiviral medications. Aztecs (AD 1100) performed medical research and recorded medicinal uses for at least 1200 plants. American Indians of Mesoamerica were the first people in the world to develop a public health care system and hospitals. They also possessed anatomical knowledge, knew orthopedic techniques, performed complicated surgeries, and could remove cataracts.

- Pharmacology: This entry shows the effective use of a great number of plant-based drugs and their proper dosage. According to the National Institute of General Medical Studies (NIGMS, part of the U.S. National Institutes of Health), the pharmaceutical wisdom of ancient cultures was impressive. According to NIGMS, the Natives of North and South America cultivated vast gardens of medicinal herbs. Daniel Moerman, foremost expert on North American Indian ethnobotany, states that North American Indians identified medicinal uses for 2564 species of plants, over 200 of which became part of the U.S. Pharmacopoeia, beginning in 1837.

- Zoned biodiversity: This is the practice of systematically planting crops in various locations in order to provide for a number of growing conditions, including soil compositions, drainage, temperature, available sunlight, and rainfall. Modern agronomists urge farmers to use this practice to prevent crop failure.

One of the most significant differences between American Indian pharmacological practices and those that evolved in Europe during the Middle Ages and demonstrates not only knowledge but also a more specific aspect of that knowledge, is that Indigenous physicians generally prescribed one plant-based medication to treat a particular condition, whereas European pharmacists ("apothecaries") used medications containing a number of ingredients. For example, Theriac, a late Renaissance medicine, contained 64 ingredients, including the skin of vipers;

it was touted as a cure for nearly every known illness but was available only to the wealthy.

Chapter
6

THE AMAZING QUANTUM
STORY

No one living in the late 19th century could have guessed the events that were about to occur in the world of physics. Close to the end of 1900 the physics climate seemed rather quiet; all of the important laws and facts of science had been discovered—so it was widely thought. Within the prior two or three decades, some important work had been done in physics. The electron, originally called a "corpuscle," was discovered by J.J. Thomson in 1897. The Michelson-Morley experiment (1887) had failed to detect an ether, disproving the ether hypothesis; instead, it proved that the speed of light was constant relative to the earth's motion. When the speed of light was computed as the earth moved in parallel with the light as well as perpendicular to it, it was the same in both cases.

James Clerk Maxwell's publication of his equations of classical electrodynamics, which describe how electric charges and electric currents act as sources for electric and magnetic fields, was an important milestone. His work also showed that the empty space in atoms was filled with electric and magnetic fields. Pieter Zeeman, a Dutch physicist, had discovered that spectral lines can be split into several components by a magnetic field—a phenomenon known as the Zeeman effect, which astronomers use to measure the strength of the magnetic field on the surface of stars; magnetic resonance imaging (MRI) is another important application.

Lord Kelvin's "clouds"

On April 27, 1900 Lord Kelvin (William Thomson), a British physicist, gave a speech to the British Association for the Advancement of Science on the topic of "dark clouds" hanging over two theories, referring metaphorically to two phenomena believed to be the only ones in physics which remained unexplained. He was expressing the confidence among physicists that all of the important laws and facts of science had been discovered; as a consequence, there didn't seem to be much future in physics. These final two problems seemed small; consequently, experienced physicists were advising potential graduate students to choose other scientific fields where they would be more likely to succeed and contribute to knowledge.

One of these two classical theories that remained unexplained was associated with light propagation, the other with heat, and physicists believed that both could be explained as modes of motion. It was also commonly believed that light waves, like sound waves, were not able to travel through space, that they required a physical medium—a hypothetical elastic stationary substance called the luminous (or luminiferous) *ether*. One of Lord Kelvin's "clouds," therefore, was the failure of the Michelson-Morley experiment conducted in 1887 to detect the supposed ether. The other "cloud" was the inability of current electromagnetic theory to predict the same bell-shaped curve (graph) that was obtained experimentally, representing light radiating from a hot object when the object was gradually heated and rose in temperature. The bell-shaped curve was the resulting graph obtained in laboratory experiments when the radiation intensity from the heated object was plotted as the frequency of the radiation increased.

Current electromagnetic theory using the Rayleigh-Jeans law or formula matched experimental results only at sufficiently high wavelengths (low frequencies) of colors. As the frequency of the light increased, which implied more intense radiation, the formula incorrectly predicted infinite power at very high frequencies (in the ultraviolet range of color), a result that was catastrophic, and the problem became known as the *ultraviolet catastrophe*.

To understand this, imagine waves crossing a given point in space. The number of waves that cross per second is called the *frequency* of the wave, and each frequency corresponds to a definite color. The wave's *amplitude* is its height above the neutral position. Conventional (classical) thought tells us that vibrations can occur to any intensity, implying that light energy can increase continuously and indefinitely as the frequency increases, but this is not true. Only certain frequencies, or amounts of energy, are actually possible.

Ironically, Lord Kelvin's "clouds" were the clues that led to the discovery of relativity and quantum physics and the birth of modern physics. In one of Einstein's published papers of 1905, he demonstrated that the hypothesis ether was superfluous; the theory he developed did not require an "absolutely stationary space." Not long before his death, he resurrected the notion of an ether, saying that space-time was in fact a type of ether. As for the puzzling ultraviolet catastrophe, it was resolved, as we shall see, by the quantum—a discrete and indivisible amount of energy.

Discovering the Quantum

The story of modern physics begins with the quantum's discovery in 1900 and coincides with the resolution to the ultraviolet catastrophe enigma. The key question that defied a correct answer under classical assumptions was "How do heated bodies radiate energy?"

All matter emits electromagnetic radiation at temperatures above absolute zero—the theoretical temperature on the Kelvin scale at which a body's molecules are at rest, implying zero kinetic energy, the thermal energy that determines temperature. Radiation represents conversion of a body's thermal energy into electromagnetic energy due to internal molecular motion and is therefore called thermal radiation. Conversely, all matter absorbs electromagnetic radiation to some degree.

It was generally thought that heat causes the molecules and atoms of a solid object to vibrate, and the molecules and atoms were themselves complicated patterns of electrical charges. It had been confirmed that oscillating charges emitted electromagnetic radiation; light, for example, was this kind of radiation. So, when a body was heated, the consequent vibrations on a molecular and atomic scale inevitably caused charge oscillations to occur, giving off heat and light. The oscillations occur as an electron moves from one orbit to another. However, this *classical* Newtonian explanation was flawed.

During this time, December 1900, German physicist Max Planck had also been trying to understand the emission of radiation from heated objects using classical assumptions. He imagined tiny oscillators acting within the black body.[27] Planck found that he could account for the bell-shaped curve obtained from observations if he required these oscillators *not* to radiate energy continuously, as the classical theory demanded, but

---

[27] The oscillations were later understood to be subatomic events. A "black body" is a hypothetically perfect, opaque, and non-reflective physical body that absorbs all incident electromagnetic radiation, regardless of frequency or angle of incidence.

The Condor and the Eagle

resonated only at certain frequencies and exchanged energy in *discrete* chunks or packets, each consisting of an integral multiple of the smallest unit of energy, and that the energy is proportional to the frequency of the oscillator. Planck relied on his intuition: If light radiates with quantized values of energy, is there a minimum unit of energy?

Newtonian theory wrongly predicted that a hot poker would give off its electromagnetic energy[28] continuously at frequencies beyond the ultraviolet. Planck's proposal was purely hypothetical, saying that the energy is emitted *discontinuously* in lumps of energy, or *quanta—a* "crackpot" idea that had no basis in a mechanical universe.

Proportionality always involves a constant, called the constant of proportionality. As an example, consider a balance typically located in the vegetable section of a store. Assuming it uses a spring, if the load F applied to it (the weight of the produce) is directly proportion to the resulting compression x of the spring ($F = kx$), then the ratio of F to x will be the same for that spring using different loads, and k is the constant of proportionality for that spring. The relationship of F to x is said to be *linear* because the "power," or degree, of the independent variable x is one. If the power of x were two (*i.e.*, x^2), the relationship would be nonlinear and the equation would be of second degree.

Planck worked with this constant and carefully calculated its value experimentally. He found that each quantum consists of energy E in the amount hf (*i.e.*, $E = hf$) for an oscillator of frequency f, where h is the *Planck constant of action*. It has the dimensions of action expressed as units of energy multiplied by time. In other words, by assuming that energy can only be absorbed or released in tiny, differential, discrete packets, which

---

[28] Light is a form of electromagnetic energy, including radiation from a hot object.

he called "bundles" or "energy elements," he accounted for the fact that certain objects change color when heated. Each packet of energy is called a *quantum*. The Plank constant applies to all microscopic events; it is a universal constant of nature.

A quantum is an indivisible unit; there are no half (½) or other fractional values of a quantum. As an analogy, imagine climbing stairs and let each step represent a quantum. You cannot take half steps or quarter steps, only whole steps or multiple steps, and each step occurs instantaneously. This kind of motion is discrete or *discontinuous*. The transfer of a quantum of energy is one of the basic events in the Universe and cannot be broken into smaller processes. A wheelchair moving up or down a ramp, on the other hand, can be thought of as the classical analogy of *continuous* energy transfer.

A New World of Reality

Quantum theory, now more than one hundred years old, immediately opened for physicists a new and surprisingly strange world that raised questions and created paradoxes which, although much has been learned, are still unresolved more than a century later. Within a decade or two after the quantum's discovery, Albert Einstein further shook the foundations of physics with the theories of relativity that changed how we think about space, time, and gravity. Together with M-theory, the quantum and relativity theories constitute what we refer to today as *modern physics*, also known as the "new" physics.[29]

Nature's surprising behavior, compared to events that occur at the macroscopic level of ordinary daily experience, is not only different from the way physicists once conceived her but is so bizarre that they had to radically change how they

---

[29] Modern physics and *modern science* (discussed in a later chapter) do not have the same meaning.

think about the real world. All of us, in fact, live in the same world of microscopic and macroscopic events. We do not live in an objective world as we once thought.

The quantum's behavior is so strange that eminent physicists like Amir Aczel refers to quantum mechanics as "the strangest field in all of science."[47] So it's not uncommon to hear a Native speaker affirm that quantum physics is now proving some mystical aspect of nature that Indians have always known. For example: "Quantum physics is proving what Indigenous peoples have known all along, that prayer can change the pollution in the waters." Also: "although quantum physics is only proving it now," moving energy through ritual has been used to heal for many thousands of years."

This book discusses in detail several specific theories and concepts in modern physics that are consistent with long-held beliefs which are widely shared among Native peoples. It goes further to discuss many strange aspects of science and the quantum world.

After the quantum, the theories of relativity soon followed and, since then, knowledge about the behavior of the Universe has grown phenomenally through ongoing research in cosmology, mainly through the efforts of theoretical physicists. Quite unexpectedly, we have learned that the cosmos is a highly interconnected and undivided whole, much like a tapestry or web. The quantum's strange behavior has also spawned various metaphysical interpretations. Modern physics has thus moved science into an area approaching mysticism. The entire Universe, a quantum Universe, as large as it is, is sustained by its smallest parts and functions because of the incessant activity occurring at the quantum scale.

These connections about the cosmos are central tenets that have been embraced by Native peoples[30] since time immemorial. They even use some of the same terminology employed by physicists. The quantum actually permeates the entire Universe, so it behooves us, especially those of us who are Indian, to affirm or re-affirm and contemplate the quantum world as an integral part of the physical and spiritual world—which I view as existing singularly together—as a part of total reality. We must acknowledge it because it's there, underlying the visible and visualizable world, and not because it's a "scientific" entity. The functioning of the universe depends on actions that occur— and are always occurring—at the quantum level.

Quantum Mechanics Timeline

A careful examination of modern physics concepts from an Indian perspective raises several questions: the validity of the whole of human experience without being limited by Western scientific considerations; whether the Universe is conscious; the nature of reality; the meaning of time and space; matter as the manifestation of energy (or spirit); physical reality in visible, invisible, and unvisualizable forms; transformations occurring at the quantum level; the relationship between energy and spirit; the implications of a whole and undivided Universe; and possibly other questions.

The story of modern physics begins, of course, with the quantum: the discovery that energy is transferred in discrete packets rather than continuously. Yet it is interesting that before the 5th century BC, even before Socrates, Greek philosophers began to speculate on key questions that are still unresolved today: What is the nature of reality? What is the ultimate constituent of the Universe?

---

[30] The terms "Native," "tribal," and "Indigenous" are often used interchangeably in this book.

In an attempt to answer these questions, the *atomists*[31] in ancient Greece believed in a mechanistic world of inert atoms and mechanical causation where there is no ruling intelligence or divine providence and no destiny or life after death. A Universe in which reality is a lifeless piece of machinery poses the *problem of free will* because a deterministic chain eliminates human responsibility. If the world is totally subject to rigid mechanical causation, then human action is a random result of atomic motions and collisions and, therefore, without freedom; and if humans cannot choose freely, they are without responsibility. Atomism also poses the difficulty of explaining cohesion, such as what keeps a rock in one piece, where does coldness come from, and where do characteristics of living things come from.

Atomism returned in the 16th and 17th centuries and was still prominent, at least in science, at the beginning of the 21st century. Physicist David Bohm has explained that the philosophy of *mechanism* is an extrapolation of Newton's laws universally applied to everything that exists in the Universe, including living systems. It is similar in concept to ancient atomism in that both are mechanistic, but the modern version is more advanced and has become scientific dogma even among some contemporary scientists.

What follows is an overview of modern physics concepts consisting of key milestones in the development of quantum theory intended to give the reader a "feeling" for modern physics and the awareness of another dimension of reality that is compatible with it. The story is told in narrative form and the discoveries are fitted into a timeline; I add my comments as each concept is introduced. Each of the cases introduces a new

---

[31] This word, is derived from the Greek *atomos* ("that which cannot be cut into smaller pieces"), which refers to an indivisible piece of matter, not to the modern-day atom.

phenomenon requiring the abandonment or modification of a classical assumption. Some topics that are introduced in this section may be re-visited later when it will become more appropriate to add details or to discuss their implications. It is worthwhile to notice, whenever possible, how an idea was inspired or intuited before it led to a new concept that was quantified mathematically.

Wave nature of light

Prior to the advent of the quantum, Thomas Young, a brilliant scientist—in fact, one of the polymaths[32] in history—had confirmed the wave nature of light since the early 19th century, contradicting Isaac Newton's long-held notion of light as a stream of emitted particles, which he called "corpuscles." Young demonstrated the phenomenon of wave interference, first with water using a ripple tank and later with light using his now-famous double-slit experiment. In his original experiment, a light beam passing through the two slits on a thin plate is observed on a screen behind the plate and, due to the wave nature of light, the waves interfere producing bright and dark bands on the screen. It is not difficult to visualize, as in the case of water waves, that when two wave fronts collide, their combined energies will form a higher crest while a collision between a wave crest and a wave trough will result in their canceling each other or forming a smaller crest.

It was confirmed beyond doubt in 1861 and 1862, when James Clerk Maxwell developed a complete wave theory of light consisting of four papers, known as Maxwell's equations of

---

[32] Young is considered polymath because of his expertise and notable contributions in a significant number of different subject areas including the fields of vision, light, solid mechanics, energy, physiology, language, musical harmony, and Egyptology. At the age of fourteen he had learned Greek and Latin and was acquainted with French, Italian, Hebrew, German, Chaldean, Syriac, Samaritan, Arabic, Persian, Turkish, and Amharic.

electromagnetism. Still a classical theory, it was a major accomplishment, no less significant than Newton's laws of motion, because it firmly established not only that light is an electromagnetic (e-m) wave but also revealed that it is a *transverse* continuous wave consisting of electric and magnetic fields in perpendicular relationship to each other and to the direction in which the wave propagates. It further proved that e-m waves can propagate in a vacuum; *i.e.*, empty space. Water waves are *longitudinal*; they move along the water's surface.

Particle nature of light

Five years after the quantum phenomenon was identified, light was again scrutinized and discovered to possess a particle nature after all. An examiner at the Patent Office in Bern, Switzerland, who read the *Annalen der Physik* scientific journal regularly, wrote mostly from the ground floor of his apartment in that city and submitted four papers to the journal, all of which were published in the same year (1905) despite not having easy access to a complete set of scientific reference materials. Those papers forever changed physics and how the world would view space, time, and matter. It was an extraordinary year for that clerk, Albert Einstein.

Each paper focused on one of the following topics: the photoelectric effect, Brownian motion, the special theory of relativity, and the equivalence of matter and energy ($E = mc^2$). I will comment next on the first, second, and fourth papers, postponing comments on the third paper (the special theory of relativity) until I can discuss it together with Einstein's paper on general relativity theory, which he published in 1916. Repeating the (long) titles of those papers would probably not be useful; instead, I will mention briefly the contribution of each.

The first of Einstein's papers was inspired by the quantum phenomenon of five years earlier. By directing a light beam onto a metal surface and observing the emission of electrons that were extracted from it—a result known as the *photoelectric effect*—he was able to confirm his intuition that light consists of individual quantum particles. But it did not invalidate the wave nature of light.

The result implied that the light beam contained or consisted of *discontinuous* particles which were bombarding the electrons, some of which in turn absorbed sufficient energy to escape. Each of these particles, later called *photons*, carried the discrete, indivisible energy $E = hf$, consistent with Planck's discovery in 1900. Whereas the energy of a classical wave depends on the wave's amplitude (so designated because the energy rides with the wave), light energy is stored in the photons, not the wave; the light's intensity does not matter. *The energy of the quanta, not the brightness ("intensity) of the source, produces the effect.*

Recall that a quantum's energy depends on the frequency (or wavelength)[33] of the constituent photons. The significance of this fact becomes clear when we compare the photon's energy with the energy of a classical wave. In an ocean wave, the amplitude depends on the wave's energy because the energy rides with the wave; the higher the amplitude, the greater the energy. This fact is intuitive and has been witnessed by everyone who has felt the longitudinal waves on a beach. An electromagnetic wave, on the other hand, derives its energy from the photons, not the amplitude, exemplifying again the quantum world's counterintuitive character.

When Einstein experimented with the photoelectric effect, he also computed the proportionality constant between energy and frequency; the constant had *the same value as the Planck*

---

[33] Wavelength, usually denoted by the Greek letter λ, is the reciprocal (multiplicative inverse) of frequency; thus higher frequencies imply shorter wavelengths.

The Condor and the Eagle

constant, even though the experiment represented a different physical situation: it involved electrons, which are particles with mass. In other words, the Planck constant of action is also a part of the energy exchange between light and the electrons that were emitted from the metal's surface.

The photoelectric effect also revealed light's "dual personality," a phenomenon known as *wave-particle duality*: that light has both a wave nature and a particle nature, but can wave and particle exist simultaneously? This question is one of the many unresolved paradoxes in modern physics—fundamental properties of the Universe that remain, to an extent, mysterious because each allows more than one possible explanation. According to the *Copenhagen Interpretation* (CI), wave and particle cannot exist simultaneously;[34] which of the two is observed depends on the experimental apparatus used for observation and measurement. Niels Bohr was the foremost advocate of CI, which is the standard, orthodox interpretation.

The late David Bohm, on the other hand, a renowned physicist and an expert in quantum theory, was among the few physicists who did not accept the CI. His last book, *The Undivided Universe*, published two weeks after his sudden death, is a complete mathematical treatise that describes his alternative theory of quantum mechanics based on an *ontological* approach, in contrast to CI's *epistemological* approach. According to Bohm's interpretation of the double-slit experiment, each particle has a well-defined trajectory and passes through exactly one slit while the wave function travels through both slits.

Bohm's passion was to move the physics of quantum theory beyond the conventional world of paradoxes (including wave function collapse) to a deeper level, a deeper order of reality. Conventional quantum theory has a method for

---

[34] There are other aspects to CI besides this claim.

computing the probabilities of experimental results but it does not give a physical account of the quantum process itself. Bohm lamented that the current theory focuses only on how knowledge of reality is obtained (epistemology) but says little or nothing about the reality itself (ontology). He introduced new terms in order to better describe the shift from the conventional epistemological interpretation of quantum theory, which focuses on the question of how we obtain our knowledge (what property is being observed?) to his new ontological interpretation, which focuses on that which exists (what is the nature of what we observe?). In Bohm's terminology, the term "observable" was replaced with "beable" (from the infinitive "to be") to indicate that, not only does the particle need to be found (observed) but it also exists even before it is observed

There is a philosophical aspect to this paradox which influences how experimental results are interpreted. Bohr was the main advocate of the principle of *complementarity*, of which the Copenhagen Interpretation is one aspect. (Some physicists say that he was influenced by the philosopher Søren Kierkegaard, who was also Danish.) Wave and particle complement each other and both are part of reality but exist separately. In contrast, in a monistic view, reality is a unified whole; there is only one world, spiritual and physical, not separate spiritual and physical worlds. Hence, light has the unified nature of both wave and particle; this coincides with my view.

Brownian motion

Einstein's second paper treated the motion of small dust particles suspended in a stationary liquid in which the visible particles are observed to move, indicating that something invisible must be causing the motion. Using a microscope, Robert Brown, a botanist, had noticed the phenomenon much earlier in 1827 with pollen grains suspended in water but could not explain it.

Einstein's paper explained the phenomenon in full detail, showing that the motion was caused by (invisible) particles in the fluid colliding with much larger dust particles. He derived statistical expressions for the displacement of these particles in the liquid (or gas) and by so doing, provided empirical evidence for the reality of the atom because the motion of dust particles was the effect of bombardment by the fast random motion of atoms or molecules in the gas or liquid. Before the paper's publication, physicists recognized atoms only as a useful concept; atoms and molecules were the subject of debate but not serious enough to argue their existence as real entities.

The molecular model consists of a volume containing a large number of constituent tiny particles which make up the fluid (gas or liquid). The particles are widely separated and free to move at random. (The distance between particles is assumed to be large compared to the size of the particles themselves.) There are no forces between the particles; the only force they experience is when a collision occurs.

Using this model, Einstein derived statistical mathematical expressions based on the kinetic theory of fluids, which at the time needed to be explained and which his paper firmly established. Kinetic theory is also how temperature is measured; temperature is a measure of the average kinetic energy in the fluid; it rises as the kinetic energy increases. (Obviously, it makes no sense to talk about the temperature of a single molecule or atom.)

Equivalence of mass and energy

As indicated above, *kinetic energy* is the energy of motion. Every object in motion possesses KE. But an object at rest also has internal energy, also known as *rest energy,* given by Einstein's famous formula, $E=mc^2$. The constituent particles within every object, even if stationary, have kinetic energy due to their violent motion. This reserve energy, or rest energy, within every object is invisible.

There is an awesome aspect about Einstein's discovery of the equivalence of matter and energy. From $E=mc^2$, where c is the speed of light and m is the mass, we can easily compute the amount of energy contained in one gram of matter (which weighs 1/28.35 of an ounce) by expressing it more meaningfully in terms of power: One microgram is 10^{-6} gram, or 10^{-9} kilogram. The speed of light is about 3×10^8 meters/second, so the energy in one microgram is obtained by multiplying mass and the speed of light: $(10^{-9})(3 \times 10^{-8})^2 = 90$ joules. If this amount of energy is delivered each second, it would equal 90 watts of power. One gram would release a million times this amount. If homes need an average of 1200 watts each, this is enough power for up to about 75,000 homes.

The nucleus

The existence of the atom was established but its structure awaited discovery. In 1909, Ernest Rutherford, already a Nobel laureate in chemistry, began to experiment with gold foil on which, under his direction, a beam of positively-charged particles was aimed. He observed that most of the particles passed through the foil while a few were deflected at large angles. Based on results, he inferred in 1911 that atoms carry a positive charge that is concentrated in a very small nucleus. He became known as the father of nuclear physics and was considered the greatest experimentalist since Michael Faraday.

The Bohr atom

In 1913, Niels Bohr, a Danish researcher who directed the Niels Bohr Institute at the University of Copenhagen, and one of the great pioneers of quantum mechanics,[35] followed up on Rutherford's concept of the nucleus located at the center of the atom. In his published model of atomic structure, known as the Bohr model, he introduced the theory of electrons traveling in circular orbits around the nucleus; when an electron dropped to a lower-energy orbit, it would emit a photon of energy. Another key feature was that the energy of the particles in the Bohr atom were restricted to certain discrete values. Rutherford referred to it as a "planetary" model, a name that was later discredited. Bohr was in the team of physicists who were assigned to the Manhattan project.

Matter acting like waves

The theory of atomic structure continued to develop as Arthur Compton experimented with X-rays, publishing his results in 1923. He observed that in a collision of an X-ray photon with an electron, the total momentum of the system was conserved and the wavelength of light changed appropriately to account for the energy difference, confirming that the photons were behaving like particles.

A Ph.D. student by the name of Louis de Broglie,[36] inspired by Einstein's work on photons, intuited that the dual nature of light might be true of *all* particles, including electrons, not just light. The electron was now believed to have wave properties. He pursued this idea in his dissertation in 1924, using the same formula, $E = hf$, which Einstein had applied to photons. Combining this formula for photon energy with Einstein's mass-energy equivalence equation, $E = mc^2$, he wrote the equations relating the momentum[37] p of a particle to its "de Broglie

---

[35] Quantum mechanics can be used interchangeably with quantum theory. It is the quantum counterpart of classical mechanics, which focuses on the physics of motion.

[36] Prince Louis de Broglie was a French scientist and an aristocrat.

[37] Momentum is defined as mass times velocity.

wavelength," typically denoted by λ (Greek letter *lambda*). The de Broglie wavelength of an object is equal to the Planck constant divided by the object's momentum:

$$\lambda = h/p \text{ or equivalently } h = \lambda p$$

This meant in particular that the electron has a wave-like property and, in fact, all particles have wave-like properties. Moreover, *every object regardless of size or mass has an associated de Broglie wave.* The existence of "matter waves" is not just a theory or an abstract idea; experiments have confirmed that objects actually behave like waves; one indicator is that they cause wave interference. This was an important result because it explained why electrons are confined to quantized orbits around the atom's nucleus; it also opened more doors to physicists who were developing quantum theory.

An examination of the de Broglie formula (in the second form above) tells us that, since the Planck constant is very small,[38] an object with a large momentum must have a short de Broglie wavelength; a small momentum implies a long wavelength.

Consider a bowling ball. Its momentum is huge, even if it moves slowly, compared to the Planck constant, which implies that its de Broglie wavelength is very *small*. Here's how: If the bowling ball's mass is one kilogram (about 2.2 pounds in weight) and it moves at the speed of one meter (about 3.2 feet) per second, its wavelength would be about 10^{-33} of a meter (a decimal point followed by 33 zeros)—extremely small compared to the size of the bowling ball itself. It is so small as to be unnoticed and its quantum effects ignored in our normative experience.

---

[38] The Planck constant's value is 6.624×10^{-34} joule-seconds

The Condor and the Eagle

However, at the molecular or atomic level, the waves begin to approximate the size of atoms and smaller particles, so the effects are very noticeable indeed, such as when an electron and a photon collide. In other words, when an electron is probed with light (a beam of photons), the electron is disturbed.

Here is a macro-level analogy: You have undoubtedly seen water waves moving against the hull of a boat. Small waves (short wavelength and amplitude) have little effect on the boat but waves that approximate or exceed the size of the boat make a difference; the boat can be buried or toppled over by the wave. An extreme case is a tsunami coming against a ship as depicted in films. Although these are not de Broglie waves, the analogy is easily envisaged. In another analogy, we can "see" an electron in the same way we can see a car—by shining light on it and having the light (photons) reflected toward our eyes (or a detecting device). Cars are much larger than the wavelength of visible light, but to observe the electron we need light (an electromagnetic wave) whose wavelength is shorter than the length of the electron's wave packet.

The de Broglie wavelength also has a very useful effect in macro-level applications. In classical physics, in order to overcome a barrier, an object must have enough energy but quantum particles, such as electrons, can take a shortcut and pass *through* a barrier if it is thin compared to the particle's de Broglie wavelength; the wavelength must extend past the barrier on each side—a phenomenon called *quantum tunneling*. Classical objects must have sufficient energy to go *over* the insulating barrier; otherwise it will recoil from it.

This tunneling phenomenon sustains ongoing processes in nature, such as nuclear fusion in stars where hydrogen atoms are converted into helium. Tunneling also has applications in millions of modern electronic devices, one of which is the tunnel diode, a kind of semiconductor diode. The diode is constructed so that two parts of the semiconductor material are separated by a very thin insulator. Electrons cannot "jump over" the insulating partition;[39] the only way for them to move across it is to tunnel *through* it.

The first semiconductor diode was invented at Bell Labs in 1939.[48] Semiconductor transistors are everywhere today. In the late 1950s, I was in charge of operations for a Bendix G-15 computer whose circuitry required 580 vacuum tubes, 27 miles of wiring, and its own refrigerated air conditioning. Later, in the early 1960s, while teaching physics, mathematics, and an electronics lab (based on vacuum tubes) at the University of Texas at El Paso, I was introduced to transistors. In 2010 the number of transistors on a single microchip exceeded one billion.[49] We can only imagine the number of computers, USB flash drives, optical drives, and other devices now in use because of the application of quantum mechanics.

I'm a user of modern technology but I still need to be reminded that nature has the intelligence on which human inventions are based. Quantum tunneling is not a human invention, it's the way nature works.

A powerful exclusion rule

In 1925, the year following de Broglie's dissertation, one of nature's most significant rules, governing how electrons configure themselves in orbits (or shells) around the nucleus of all atoms, was discovered by Wolfgang Pauli. According to the rule, no two electrons can be in the same quantum state simultaneously. Known as the *Pauli Exclusion Principle*, the rule can be stated in general terms as follows: No two identical *fermions* can be in the same quantum state simultaneously.

---

[39] In quantum mechanics, a barrier is often depicted as a wall, even though the physical barrier is not a wall.

The Condor and the Eagle

A fermion is a particle whose *spin*[40] is a half integer (for example, -1/2 or +1/2); electrons, protons, and neutrons are fermions. Particles with whole-integer spin, such as 0, 1, and 2, are called *bosons*. For this rule to apply, the two particles must be so close to each other that their de Broglie waves overlap. Before the exclusion rule was known, the periodic table of elements (atoms) could not be explained; in fact, without the rule, the table would not exist.

Fundamental limit on measurements

Because the de Broglie formula connects the electron's wavelength to its momentum, the wavelength limits the accuracy with which an electron can be located in the atom. There is a fundamental limit to the accuracy with which the position and momentum (or other measurable properties) of a quantum particle can be known simultaneously. Measurement of a particle's properties at the quantum scale requires a measuring apparatus.

Accurate measurements of position or momentum can be performed separately but *not simultaneously*. Werner Heisenberg in 1927 worked out the mathematical relationship, known as the *Heisenberg Uncertainty Principle*, which expresses this limit:

$$\sigma_x \sigma_p \geq \frac{h}{4\pi} \quad \text{or} \quad \Delta_x \Delta_p \geq \frac{h}{4\pi}$$

where the "σ" and "Δ" symbols denote *standard deviations* representing the range of uncertainty[41] in the measured values

---

[40] *Spin* is a magnetic-like property of a particle whose possible values are "up" and "down." It is a vector quantity; *i.e.*, it has magnitude and direction.

[41] An "uncertainty," in terms of the standard deviation, denoted by the Δ (delta) symbol, is a range of values from the lowest to the highest that contains the correct value. The standard deviation is a statistical measure of how closely

of position x and momentum p. The relationships imply that the more accurately the momentum is known, the less accurately the position is known, and vice versa. Simply stated, the product of uncertainties in position and momentum can never be less than h/4π. Here again, the Planck constant h plays an essential role.

Matrix mechanics

The emission spectrum for an atom shows the possible energy levels for that atom. Each atom has its own unique spectrum, like an identifying fingerprint. The spectral lines of hydrogen—the simplest element, consisting of only one proton in the nucleus and one orbiting electron—correspond to discrete and sudden changes in the electron's energy levels. When the energy level jumps from a higher energy to a lower one, a photon of a specific wavelength is emitted.

Heisenberg tried for some time to explain the spectral lines of hydrogen; eventually, using a series of mathematical analogies, he wrote out an equation that is analogous to the classical computation of intensities but represents instead the quantum mechanical case (which takes the Uncertainty Principle into account). He used a rule of multiplication that works for matrices (rectangular arrays of numbers) in mathematics. His matrix method was the first version of quantum mechanics; it calculates the probabilities of transitions between different energy levels. Erwin Schrodinger later developed the Schrödinger wave equation, which was shown to be equivalent.

---

grouped or how widely spaced (spread out) from the average value the data in a collection appear to be.

The observer and the observed

The distinction between the physicist and the microscopic quantum system becomes blurred in quantum mechanics. In classical mechanics, both the physicist as the observer and the system studied and measured are at the macro level, but in quantum mechanics, the physicist is at the macro level and the quantum system exists in the micro-world, the world of the quantum. These are two different kinds of relationship and two different worlds of reality, one is visible and the other is invisible. As a participant, the physicist cannot step outside of the total system and become an objective observer.

The ability to measure or detect particle properties became difficult in the following way: Since the quantum system is at the micro level, the measurement must be performed by a non-human measuring apparatus, which lies at the macro scale (same as the observer) and brings results back to that scale; however, the object being measured at the micro scale is inevitably disturbed, which affects the measurement. In quantum mechanics, the quantum system and the apparatus interact at the boundary between the two levels. But where is the boundary line?

Whereas classical physics is a science of objectivity, quantum mechanics obliges the physicist to interact with nature because the observer and the observed exist in separate worlds—the quantum and the classical, which are connected by the indeterminate boundary (mentioned in the last paragraph) that divides them—and must communicate in order to yield a measurement.

To Bohm, the observer-observed question was important because of its relevance to the meaning of quantum theory, which was central to his work, but *it was also important outside of quantum theory*—yet both aspects must be somehow connected. When he became aware of Jiddu Krishnamurti, the renowned spiritual teacher, who had written about this question from a philosophical perspective outside of quantum theory, after reading his book, *The First and the Last Freedom*, he was determined to meet him. The meeting led to a friendship and collaboration that lasted until Bohm's passing in 1992. For 25 years they held dialogues around deep but practical questions about the nature of thought and consciousness and the nature of existence. In *The Ending of Time*, they ask about the wrong turn humanity has taken and whether there is an escape.

Problems associated with quantum measurements have spawned other philosophical questions. How do we know a particle is real (*i.e.*, it actually exists) if we can only see its traces on film? How do we know that those traces identify the particle in question? Do things exist independently of being observed? Such questions can affect the meaning of science. The positivist holds that only authentic knowledge is real, which is possible only through the scientific method; a *logical positivist* would say that observational (empirical) evidence is not only necessary but it must also be verified.

According to some physicists, Niels Bohr leaned toward positivism, although we can't be sure because his explanations are said to be confusing and obscure. He adamantly held the view that physics must be confined to phenomena that can be experienced and, furthermore, that the language of physics must reflect this view; nothing should be called "real" unless it is observed. "There *is* no quantum world, only an abstract physical description of it." So it is wrong, he insisted, to think that the task of physics is to find out how nature *is*; physics is concerned with what we can *say* about nature.[50] Einstein, on the other hand, held the view that things exist before discovery.

When Heisenberg first tried to explain the Uncertainty Principle to Bohr within the old classical worldview, using terms that implied that particles *possessed* certain properties,

Bohr retorted that the electron does not *possess* intrinsic properties such as position and momentum; this assumption is a "hangover" from old ways of thinking, he remarked.

This problem is not only philosophical; it leads to ambiguity because, although terms like *wave, particle, momentum, position,* and *trajectory* have precise meanings in Newtonian mechanics, they no longer have the same meaning in quantum mechanics. According to the phenomenon known as wave-particle duality, a particle can also behave like a wave, as stated earlier. Since it has position whereas a wave does not, which term (wave or particle) is correct?

Heisenberg thought of a particle as possessing well-defined properties, including position, momentum, and trajectory, even though they could not be exactly determined—a view that provoked Bohr, who adamantly insisted that the precise path of a particle should not be called "uncertain" because the term implies that it existed before it could be determined. Rather, he said, it should be called *ambiguous,* just as temperature is inherently ambiguous because it is a measure of the amount of kinetic energy of an ensemble of molecules and has no meaning for individual molecules. The problem, he said, lies in the fact that the entire phenomenon (quantum system plus measuring apparatus) cannot be analyzed at the microscopic level of detail; observer and observed must be treated as a single system. In the same manner, he argued, it makes no sense to talk about position, particle, or momentum as if they were real entities.

In his old age Heisenberg expressed the view that any talk about electrons and protons as "building blocks" of matter was a confusing misrepresentation of the nature of quantum reality. Rather, he said, these entities are the *surface manifestations* of underlying quantum processes; symmetries (such as dualities) are the more fundamental property. Reality lies not with the particles but with the *relationships* that exist within the flux of energy and processes of quantum nature.

Kafatos and Nadeau (*The Conscious Universe*) offer a perspective about the observer/observed relationship that involves the physicist's mind and consciousness:

> In quantum physics, observational conditions and results are such that we cannot assume that there is a categorical distinction between the observer and the observing apparatus or between the mind of the physicist and the results of physical experiments. ...we can no longer see the pre-existent truths of physical reality through the lenses of physical theory in the classical sense.[51]

They also state that, since by necessity the act of observation involves both the measuring apparatus and the observer, the truths of science must be treated subjectively. Slowly but surely, this healthy reasoning that has occurred among physicists seems to corroborate the importance of relationships and subjectivity—a process that approaches an Indigenous perspective.

The wave equation

Heisenberg had already developed the first version of quantum mechanics that made use of matrices to calculate the probabilities of transitions between energy levels. There was no equation which described the behavior of a quantum mechanical wave; *i.e.,* the state of the quantum system in terms of *fields, interactions,* and *transformations,* instead of classical quantities (position, force, and momentum) for individual particles.

Newton's second law of motion in its simplest non-relativistic mathematical form (stated without calculus) is $F = ma$, where F is the force acting on a body with mass m, and a is the acceleration resulting from the applied force. To apply this law to the trillions of trillions of molecules in a room is virtually impossible. Physicists are aware that working with a two-body problem can become difficult and the difficulties of working with the three-body problem increase dramatically. It involves taking an initial set of data specifying the positions, masses and velocities of three bodies at a particular instant in time and predicting their motions at a later time, using Newton's laws of motion and gravitation. We are only dealing with classical cases; in the story of development, the gravitational force has yet to be incorporated into quantum theory.[42]

The quantum wave equation must take matter waves into account. Erwin Schrödinger, inspired by de Broglie's insight regarding matter waves, took on this task in 1926 and in six months succeeded in deriving an equation. After several false starts (because he was initially using a classical approach), he derived the equation using the following quantum relationships which have already been stated:

$$\lambda = h/p$$

$$E = hf$$

$$\Delta_x \Delta_p \geq \frac{h}{4\pi}$$

Recall that the first equation above relates a particle's de Broglie wavelength to its momentum, the second one relates quantum energy to the frequency of the wave, and the third is the Heisenberg Uncertainty Principle. Electrons move in orbits, of course, but the above equations can easily be converted into terms of angular frequency and angular momentum.

---

[42] Physicists have worked for decades on the problem of combining gravity and quantum theory into a unified theory of quantum gravity.

The resulting Schrödinger equation in its simplest one-dimensional, time-dependent form, representing an electron moving along a straight line (in the x direction) is

$$i\hbar\frac{\partial}{\partial t}\Psi = -\frac{\hbar^2}{2m}\frac{\partial^2}{\partial x^2}\Psi + V(x,t)\Psi$$

The right-hand side of the equation expresses the total energy. Ψ (the Greek letter psi), called the *wave function*, is the dependent variable whose value we want to determine. It represents the probability of finding the particle at position x; m is the particle's mass; $V(x)$ is the potential energy of the system; and $\hbar$ (pronounced "h-bar") has the value h/2π and contains the familiar Planck constant. The symbol "i" is called the "imaginary unit," a constant equal to the square root of -1. On the right-hand side, the expression containing the "∂"symbol, designating a "partial derivative" in mathematics, gives the rate of change of the wave function Ψ as the curvature of space changes (but in the one-dimensional case, there is no curvature). On the left side, the expression containing the "∂"symbol gives the time rate at which Ψ is changing.

What is the meaning of Ψ? Schrödinger originally thought his equation was about (real) matter waves and that it predicted the position of an orbital electron. But in the same year (1926), Max Born demonstrated that the wave function, Ψ, is unobservable and that its square ($|\Psi|^2$) gives the *probability* of finding the electron (or other particle) within a certain region of space. Interpreted this way, the Schrödinger equation gives probabilities that matched experimental results. This outcome turned out to be consistent with Heisenberg's Uncertainty Principle. In addition, Schrödinger subsequently demonstrated that his equation was mathematically equivalent to the matrix

The Condor and the Eagle

method of quantum mechanics that Heisenberg had derived earlier.

An issue that has been heavily debated by physicists is the "collapse of the wave function." The wave function provides a complete description of the quantum particle or system; however, at the time of measurement, only one possibility actually materializes. Here is how Kafatos and Nadeau describe it:

> "What this formalism indicates is that prior to measurement we only have a range of possibilities, given by the wave function, in terms of mathematically derivable probabilities, given by the square of the amplitude, $|\psi|^2$. When an actual measurement is made, or when something "definite" is recorded by our instruments, the various possibilities become one "actuality."[52]

What would happen if, when the wave function collapses, instead of falling on only *one* of the several possibilities defined by the function, *all* of the possibilities are actualized? This is the "many worlds" interpretation of quantum mechanics. If this kind of collapse happened at the Big Bang, it could have generated multiple, parallel, and independent universes, of which ours is one.

The Schrödinger equation satisfies two crucial requirements. First, although it describes a continuous wave, the continuity relates to the range of probabilities, not to the quantum actions themselves. Second is the fact that the equation takes the Uncertainty Principle into account. But how can a continuous wave represent discrete quantum phenomena, considering that quantum mechanical measurements cannot be performed simultaneously with unlimited accuracy? Quantum theory represents a radical change in our fundamental conceptual framework; it implies a renunciation of the manner in which we imagine reality. These concepts are of a totally different nature from those in classical theory. It is only at the quantum level that this radical change is experienced.

I mention one final issue. Many have heard of Schrödinger's cat paradox, which Schrödinger himself created to illustrate a philosophical problem that troubled him about the Copenhagen Interpretation (and the wave function collapse associated with it) because it seemed to contradict common sense. A cat, a flask of poison, and a radioactive source are placed in a sealed box. If an internal monitor detects radioactivity, the flask is shattered, releasing the poison, which kills the cat. (There is a 50% chance of this happening.) The Copenhagen interpretation implies that after a while, the cat is simultaneously alive and dead. Yet, when we look in the box, we see the cat either alive *or* dead, not both alive *and* dead.

Some popular books equate wave function collapses with human brain events: when someone makes a decision, it causes a wave function to collapse in the brain, an idea that connects human consciousness with quantum theory. Since public intrigue with quantum theory is not free of misconceptions, it's important to be somewhat skeptical about what is being claimed. One of the claims is that quantum theory explains human consciousness. In the first place, the "collapse" is not necessarily a real phenomenon; it is one of the ways in which physicists have interpreted quantum mechanics; indeed, it relates to the standard interpretation. Other authors who are not physicists or scientists may have taken this idea (of equating quantum collapses with brain events) from books authored by physicists or perhaps conceived it independently after reading about quantum theory. In any case, it must be remembered that the quantum event, if it actually occurs, is just that: an event in the laboratory which, in my view, has little to do with human consciousness. On the other hand, what happens in the brain when we make a decision may be a quantum event; however, to complete this argument, one would still need to show that quantum events are linked directly to quantum function collapses.

Non-locality

Quantum theory predicts a phenomenon that so strange that Einstein himself would not accept it, calling it "spooky." But despite his attempts to discredit it, numerous experiments conducted after his death vindicated the predictions of the theory and proved him wrong.

In classical thought, an object **A** cannot have an instantaneous effect on a remote object **B** because the two particles are independent of each other; **B** can be influenced by **A** only by forces or signals within reach in its local environment, such as a gravitational force or by receiving signals from **A**; these are influences that take time to arrive and they cannot arrive sooner than the speed of light will allow. But in quantum mechanics, two particles can become *entangled* so that, when the property of particle **A** undergoes a change, the other, distant particle **B** will instantaneously undergo a corresponding change in that same property, regardless of the spatial distance between them; *i.e.*, **B** alters its property so that the quantum state of the two particles remains unchanged. This phenomenon is known as *non-locality*.

For example, consider the spin property of the particles. When **B** is observed to be the "spin up" state, **A** will change to a "spin down" state, and conversely. Note that the spin states are not known in advance; in this example, when a property of particle **B** is measured, the corresponding change immediately occurs in the other particle.[43] This implies that only one

---

[43] Since spin is a vector quantity with three directional components, only one component at a time can be measured.

measurement is needed, since particle **A**'s spin property can be predicted. Different versions of this experiment have been carried out successfully numerous times and with more than two particles. In a 1997 experiment the two particles were 27 kilometers apart.

Consider an analogy outside of the physics lab that contains all the elements which would appear in a lab experiment. Suppose that, in order to avoid wearing mismatching socks, you keep only two pairs in your dresser, one black and one brown. If you take the brown pair out of the drawer, you know for sure that the other one is black. Likewise, if you take out the black pair, you know the brown one is in the drawer. However, these are not ordinary socks; they have the following magical property. Neither pair reveals its color until it is exposed to light. Each is kept inside a box; when the box is opened, it will reveal the color black or brown, and there is a 50 percent chance that either color will appear. If the color is black, you will know for sure that the other pair is brown, and vice versa.

You now go on a trip, picking a box from the drawer that contains one pair of socks and taking it with you. You will not know its color, of course, until you open the box. Eventually, you decide to open it on the trip, either in flight or at the hotel. If the pair of socks is black, the twin pair at home *instantaneously* changes its color to brown. If the socks show a brown color, the pair at home will be black. It's as if an instant message were being sent to the pair at home, telling it to change to the opposite color in order to preserve the combined brown-black state.

But no such message or signal is sent to the pair at home, nor is that possible because the change occurs instantaneously yet nothing can travel faster than the speed of light. How could the socks at home know which color was opened far away? The linked particle "knows" the color of the first particle instantaneously, regardless of distance; no signal sent to it except at infinite speed from far or near would arrive fast enough. In contrast, *locality* is the common-sense principle of classical physics; a measurement taken at one point in space cannot influence what occurs at another point beyond a fairly short distance. An object can be influenced directly only by its immediate surroundings.

Ordinarily, distant objects cannot influence one another directly; an object can be influenced only by its immediate surroundings or by communicating through signals. But it can happen if two quantum particles are entangled, violating the classical and intuitive principle of locality. It is as if the two consisted of a *single entity*.

Non-locality has broad implications. It implies *wholeness*, a concept that is often expressed as "all things are connected." It is not an esoteric or trick event that occurs only in complicated laboratory setups, but a property of the Universe which must be viewed as an interconnected and interdependent whole. It raises the possibility that, no matter where we are, what happens around us can be affected immediately by events in other parts of the Universe. Quantum theory calls for a complete change in worldview to a perspective that is consistent with how Native peoples relate to the world. The previously held principle of *locality*—that the Universe consists of entities that can be considered separate from each other and influenced only by their local surroundings—is fundamentally wrong. Some physicists suggest that all matter became entangled during the Big Bang. If all quanta in the Universe are interconnected, what does that say about our consciousness of the whole cosmos?

Quantum Field Theory

Relativistic effects due to high-speed electrons needed to be incorporated into the Schrödinger wave equation. It was Paul Dirac, a young British physicist of 26, who took the important step in 1927 of bringing quantum physics into conformity with Einstein's special theory of relativity. He developed an equation that describes the behavior of electrons at any speed up to the speed of light. Now called the Dirac equation, it explains one of the electron's (and fermions in general) intrinsic properties: *spin*.

It was said of Dirac that he was a man of few words who avoided publicity and celebrity status. He was considered by eminent physicists, including Einstein and Bohr, as an outstanding genius in the history of physics. According to Graham Farmelo, adjunct professor of physics at Northeastern University, Dirac was a strange man, in fact the "strangest man" and a "mystic," a description that Farmelo chose as part of the title of his book about Dirac because of the latter's "mystical" approach to theoretical physics. For example, when Dirac learned that physicists Pascual Jordan and Eugene Wigner had developed a field theory that complemented Dirac's theory, he was not impressed, saying that he could not see how the symbols they used corresponded to "things going on in nature." But he made the mistake of misinterpreting their work due to his different approach to physics: if he could not visualize a theory, he tended to ignore it.[53]

His equation had serious problems but he had "mystical" faith, Farmelo tells us. The conventional, bottom-up method is to carry out experiments and then try to find theories to explain the results. But Dirac was a top-down thinker; he first formed an eloquent equation that united quantum mechanics with relativity to describe the relativistic electron. Then, with no experimental clues to guide him, he used his equation to predict the existence of *antimatter*, previously unheard-of particles with the same mass as the corresponding particles of matter but with an opposite charge. Despite pressure from other physicists to abandon that approach, Dirac was defiant and gave priority to the mathematical form of his equation. He sought a way to interpret it, feeling that by continuing to work with it he would find the right theory to fit it, and his persistence paid off.

In 1931, he correctly predicted the existence of a particle with the same mass as an electron but with opposite charge and spin which he called an "anti-electron," now known as the positive electron, or *positron*. In 1932, Carl Anderson, a professor at California Institute of Technology, discovered the positron while measuring cosmic rays in a Wilson chamber experiment. The Dirac equation was one of the most outstanding triumphs of theoretical physics. More *antiparticles* were later discovered and Dirac became recognized as the discoverer of antimatter; in 1933, he shared the Nobel Prize with Schrödinger.

Fundamental Forces of Nature

What keeps the nucleus together, the electrons bonded to the nucleus, and the cosmos moving in an orderly manner? How do particles not in direct physical contact with each other influence each other? The answer to both of these questions involves forces.

Consider a point in space. In classical physics, the presence of a particle in space creates a field around the particle that influences other particles within range. When a second particle enters the field, it experiences a force from the first particle and both experience the same force. Because in classical mechanics the force is assumed to act instantaneously, the field effect is called "action at a distance." An example of this field effect is the gravitational field around the earth that attracts objects toward the earth; every object inside the field feels the effect. Although the earth is also attracted to each object by an equal but opposite force, it doesn't "feel" it because of its comparatively large mass.

In quantum field theory, on the other hand, particles are not "acted on" by forces as in classical physics; instead, an "interaction" or "interactive force" is produced when they exchange the quanta of other particles called "force carriers." The force carriers are also referred to as "messenger quanta" or "mediators." For example, the electromagnetic force is carried by photons; the force is produced by exchanging photons between electrons. Two charged particles interact through the exchange of photons. The photons don't transmit the force; they transmit a message, such as "come together" (for oppositely-charged particles) or "move apart" (for like-charged particles).

There are four known fundamental interactions, or forces, in nature: *strong, electromagnetic, weak nuclear*, and *gravitational* (the weakest). The strong nuclear force holds the nucleus (consisting of protons and neutrons) together; it is mediated by *gluons*. The protons have a positive charge and the neutrons are neutral (but slightly heavier than protons). The electromagnetic force, which is weaker than the strong nuclear force, keeps the electrons bound to the nucleus due to the attractive force between the positively-charged protons in the nucleus and the orbital electrons. The weak nuclear force, which is weaker than electromagnetic force, is responsible for radioactivity and the nuclear reactions in stars; it is mediated, or carried, by "W" and "Z" bosons.[44]

The gravitational force is the weakest force in nature; according to particle theory, it is *hypothetically* mediated by "gravitons"—predicted by the *standard model* (SM) of particle physics but, so far, undiscovered. Note that the strong force must overcome the electromagnetic forces of repulsion which

---

[44] "W" bosons can have a positive or negative electric charge; the "Z" boson is neutral.

The Condor and the Eagle

tend to separate the protons in the nucleus; the presence of neutrons helps to weaken the repulsive forces by increasing the distances between nucleons. The SM is based on experimental results and encapsulates the best understanding of how twelve particles and three of the four fundamental interactions (forces) are related to each other.

Consciousness and Intelligence

The Western World, including Western science, has believed historically that everything can be reduced to lifeless objects which are seemingly unrelated to each other. It perceives a dead universe, a world without spirit or consciousness, and despite the frequent references to "Mother Nature," it conceives of nature as inanimate.

Recently, the concept of "deep ecology" was introduced, which ironically is purported to be "a new way to think about our relationship to the earth" and a "complete overhaul" of the way humans live on the earth. But it is *not new* to traditional tribal peoples who have long lived in deference to the Universe, respecting its authoritative system of laws and principles and recognizing the complex inter-relationships between all things, including a proper relationship to the earth.

The role of human consciousness and the "possibility" that nature exhibits conscious behavior are issues that the Western mind wants to explore. But not only is it strange that anyone would question nature's consciousness, in view of its evident intelligence, as exemplified by how well natural processes function without the aid of human intelligence, knowledge, or muscle, but there is also no thought or desire to build a respectful relationship with nature. Any being that possesses intelligence, it seems to me, is a conscious being. From a Western perspective, the suggestion of a conscious Universe is a ridiculous and alien idea—that is, unless it is supported by empirical data so as to be considered *legitimate* science, and even then it might be accompanied by a carefully constructed caveat that disavows anything that could be construed as belief in a sentient Universe. Quantum physics has given us new knowledge which reveals that the connection between mind and nature is far more intimate than previously imagined.

There are noticeable signs that the Western paradigm may be shifting, as the titles of new or recent scholarly books and book reviews employing such expressions as "cosmic web of life," "a new vision," "spirituality," and "inter-relationship," seem to reveal. An article by Duane Elgin, for example, begins with the following statement about the co-edited book, *Scientific Evidence of a Living Conscious Universe*, by Trish Pfeiffer and others:

> I believe that the most far-reaching trend of our times is an emerging shift in our shared view of the universe – from thinking of it as dead to experiencing it as alive. In regarding the universe as alive and ourselves as continuously sustained within that aliveness, we see that we are intimately related to everything that exists.[54]

Typically, authors cite the works of previous authors from various branches of science, so there does seem to be a trend toward a new vision or a new physics. It is even more notable when eminent physicists like Freeman Dyson or David Bohm boldly affirm that the Universe is capable of consciousness. In another article, Elgin writes:

> Consciousness, or a capacity for knowing, is basic to life. If the universe is alive, we should therefore find evidence of some form of consciousness operating at every level -- and that is exactly what we find. The respected physicist Freeman Dyson writes this about consciousness at the quantum level: "Matter in quantum mechanics is not an inert substance but an active agent,

constantly making choices between alternative possibilities. . . . It appears that mind, as manifested by the capacity to make choices, is to some extent inherent in every electron." This does **not** mean that an atom has the same consciousness as a human being, but rather that an atom has a reflective capacity appropriate to its form and function.[55]

The late David Bohm was a highly respected and renowned plasma and particle physicist who is still referenced extensively in physics literature for multiple reasons—his work in quantum mechanics, his theory of the implicate order, as a compassionate philosopher, his work in plasma physics, and more. A plasma is virtually a gas comprised mostly of free particles; *i.e.*, particles that are not bound to the nucleus of atoms or any particles but exist freely in space. In his plasma experiments at the Berkeley Radiation Laboratory, Bohm found that individual electrons act as part of an interconnected whole, self-regulating as if they were inherently intelligent, and seem to have a *collective consciousness*, which implies that individual consciousness must primarily exist *within the electrons*. This aspect of his work, including his conclusions, has been widely referenced.

In a *CNN Belief Blog* article, "Science and Spirituality should be Friends," Deepak Chopra, who has authored 60 books, questions the presumed split between the quantum micro world and the visible macro world: "Either atoms and molecules are smart, or something makes them smart. That something, I believe, will come down to a conscious universe."[56]

There is an important difference, I believe, which needs to be stressed between how a traditional Indian relates to nature and how a scientist who is committed to a Western worldview relates to his or her work. In consciousness studies (at least the ones I have seen) conducted within a Western context, researchers perceive the phenomena under investigation as *objects*, referring to them in third-person language and implying that a personal relationship with nature is not within their worldview.

Studies in Consciousness

Quantum theory is partly responsible for the growing interest and activity in consciousness research among physicists and, in particular, questions about the nature of reality and its relationship to consciousness. This was one of the issues discussed in an international symposium on consciousness held in Athens, Greece in 1992. According to the online report, "The Challenge of Consciousness Research," the purpose of the symposium was to explore some of the important issues about consciousness, including how it should be studied, the adequacy of conventional methods of science, constraints on the kinds of investigations that may be conducted under the scientific method, the role of intuition, and more.[57]

Another theme discussed, according to the report, was the connection between the scientist, the science under study, the influence on society and the environment, and moral issues that cannot be separated from science. Symposium participants generally agreed on the following points:

- Extend science to include insights that may be gained from spiritual practices that involve consciousness in its many dimensions (symbols, myths, metaphors, etc.);

- attempt to overcome the many artificial dualities, such as mind and body, mind and matter, feminine and masculine, and philosophy and science;

- move from fragmentation to integration and broaden the notion of causality to account for networks and non-local connectedness;

- foster a radically different attitude of humility ("awe, wonder, and delight in the cosmos which is the beginning of all science");

- acknowledge the wholeness of body and mind (the total person) and thus avoid the Cartesian split;

- embrace both science and conscience.

The above is only a partial list of proposed extensions to science as it is currently practiced. They are *strikingly similar* or equivalent to the characteristics of a holistic science that Indigenous peoples have been practicing *as a way of life* for hundreds of years, even millennia. Traditional Indians practice these extensions to science insofar as they are permitted to and contingent upon an adequate land base. But consciousness research does not fit their agenda, since they already relate to a live and conscious Universe.

The interest in consciousness studies is one case among others that exemplifies how Western scientists are finally beginning to question some long-held assumptions of their culture and to initiate research into phenomena with the aim of acquiring new insights that involve opening the mind to unconventional ways of thinking. It is commendable for them to take the risk of being viewed as mavericks and losing their professional standing before their peers.

Recent research is challenging many of the materialist assumptions of 19th-century physics but, as mentioned earlier, it also raises philosophical issues about reality. One view among physicists, known as *metaphysical realism*, is that physical reality exists independently of human observers or acts of observation; *i.e.*, independently of the (human) mind. Some challenge this view by asking how knowledge of a mind-independent world can be obtained and how our beliefs can be linked to the properties of mind-independent quantum particles—an argument which assumes the existence of a mind-independent world. Another view, *epistemological realism*, requires strict adherence to and regard for rules and procedures for doing science; thus, no phenomenon can be considered real until it is observed. This view doesn't hold because, although non-locality, for example, is not an *observed* phenomenon, it is real on the basis of observed *correlations*. In this case, at least, the positivist view has to be abandoned.

THE THEORIES OF RELATIVITY

E instein's discovery and development of relativity theory brought radical changes to notions about space, time, matter, energy, and gravity. Classical notions about all of these were firmly established in the minds of physicists, only to be rendered obsolete seemingly in a moment's time. According to the theory, instead of objects "attracting" each other, space—or, more correctly, space-time—bends in the presence of a massive object, causing less massive objects to fall into a spatial depression. Space and time are no longer independent entities; they have been replaced by flexible space-time.

The stunning discoveries in quantum theory and relativity theory in the early 1900s completely overturned concepts that had been accepted as truth for thousands of years. Yet despite their philosophical significance, these discoveries had no noticeable effect on the worldview of mass society—a change that should have led to a paradigm shift toward a holistic view of the Universe, at least among physicists. Instead, even today, most physicists still seem to be driven by a Newtonian consciousness and vision that contradicts relativity, even though the latter has been incorporated into the mathematical framework of fundamental physics.[58]

We live and think in a Newtonian world by habit because it is the world of direct observation, accessible through our senses. But this behavior can only provide a naïve view of reality, one that is incompatible with modern physics, as eminent physicists Leonard Mlodinow and Stephen Hawking (*The Grand Design*) correctly point out.[59] Think of what this implies: With the discovery of relativity, classical physics is no longer a valid foundational principle, yet the new physics (quantum and relativity theories) *has as yet not replaced Newtonian thinking* as the legitimate central foundation either in physics or in our everyday lives. Let's state this another way: Relativity concepts have not been grafted into the *tacit infrastructure* of practicing physicists. Recall that this is what causes us to act (or react) automatically without asking ourselves why we do it. If physicists haven't converted over, much less can we expect the rest of society to do it. And this is my main point: that holistic ideas need to be implanted according to which nature herself has been revealing to us through modern physics; holism should become a part of our tacit infrastructure if we are to expect the current paradigm to change.

Space and Time

We will consider the classical notions of space and time before discussing them in the context of relativity. Classical physics deals with the world of common experience, a world ruled by common sense, where time and space are absolute. Whatever is true or real exists independently of other events and it possesses its own nature. Isaac Newton believed that time and space are real and absolute entities which are mutually independent. He also believed that time and space exist whether or not they are observed and whether or not an observer or a potential observer is present. On the other hand, when something is *relatively* true or correct, it is so by comparison to something else.

In classical thought, absolute space is the entire region of the Universe; it is where matter is thought to reside, the background where all events occur, the once-empty, inert "holding tank" that could exist on its own even if nothing were in it. It is separate from the matter it contains and is considered to have pre-existed before matter was put in it.

Absolute time is said to "flow" at the same rate everywhere: on the earth, the planets, the galaxies, and in all contexts and scales. It is as though a universal clock were constantly ticking at the same rate and recording events in space. Regardless of where clocks are located and whether those locations are moving or stationary, so long as they are synchronized, they will record the same time and tick at the same rate everywhere.

These bold notions of time and space were invalidated by the discovery of relativity. Common experience is an illusion insofar as we interpret events on the basis of absolute time and space.

Time in physics is measured with a clock, which is an ideal periodic process that consists of ticks occurring at regular intervals. But the clock measures time *interval*, not absolute time, and in mathematical equations it is typically represented by the letter "t." Duration is the number of ticks that occur between events. Any device that does this can be called a "clock" but the clocks we buy at the store are calibrated to measure time in seconds.[45]

The term "arrow of time" points to the tendency in nature to change in the direction of increased randomness instead of order. This asymmetry in nature in which physical systems change naturally and irreversibly toward increasing disorder is called *entropy*. Rather than state the thermodynamic definition for entropy, an example or two that demonstrate the principle

---

[45] The *second* is a standard international unit precisely measured in atomic terms.

will serve our purpose. The main idea is that in natural processes involving the conversion of energy from one form into another, some of the energy is wasted as unusable heat. Examples of entropy, demonstrating irreversible behavior (asymmetry), can easily be cited: an egg that is dropped and shatters into smaller pieces, the biological decay process, ice cubes melting in a glass, and so on.

Now consider time and change. It is common to think of history as comprised of a sequence of events occurring along a straight time line in causal order. History is commonly perceived from the perspective of a given location, and there are many histories. But another way of thinking is to focus on change as constantly occurring everywhere at once. That is, instead of thinking of time, think of change in every case; this is the concept of flux. Would it be necessary to consider time if change did not occur? Note that time cannot be defined independently of change. How fast does time flow?

"Flux" in physics usually means that something is flowing across a surface. But to ancient cultures from circa 500 BC, the concept of a *primal flux* was a sophisticated idea. For example, the conversion of cream into butter, in which all component ingredients were involved in a transmutative process, was likened to constant flux.

Einstein did not believe in the existence of objective time; the concept of space, he said, is likewise useful but not indispensable. He wrote once to the family of a lifelong friend who had died, stating that as a physicist his belief that the separation between past, present, and future is "only an illusion, although a convincing one." It is my understanding that time does not exist in some Native languages.

In relativity, space is never used as a reference for motion; motion does not occur relative to space but relative to objects. For example, a car moves relative to the road and the earth moves relative to the sun but also relative to the stars and our galaxy. The speed of an object has no meaning unless it is compared to other objects. All motion is relative, but that's not all, as we shall see.

Movement is also related to change or flux, according to author Joseph Rael from Picuris Pueblo, who tells us (*In the House of Shattering Light*) that in his Tiwa culture there is the idea that "we do not exist" because we are changing from one small "moment" to the next. "Perceptual living is a statement of impermanence, therefore always changing. It's as if light (Universal Intelligence) is constantly shattering into itself as it moves along themes of creativity."[60] These timeless movements and change correspond to flux and change in nature as described by relativity and quantum theory.

In the classical world, physical space serves as the background where all events occur. Indian reality is "non-classical" in that we interact with the unseen spiritual world and participate interactively with the powers of the Universe, involving the whole being. Spiritual experiences, visions, dreams, and ceremonies cannot be reduced to objects in three-dimensional space and classical time. The sweat lodge is an example; its structure has different meanings among tribes. I have been in sweats conducted by different tribes: Chippewa/Ojibwe, Nez Perce, Assiniboine & Sioux, Yakama, and others. The earth mound, the fire pit, the hole for the stones, the different shapes, structure, and objects inside and outside, the door direction—all of these have definite meanings, involve prayers, and in most cases are a sacred journey to the Grandfather. They involve symbolism but the experience is real. We participate in the flow and harmony of the cosmos, contemplating and paying deference to the Power and powers behind every movement but we don't worship the objects. On one occasion, as the Sun Dance was about to end, the last forty hot, glowing stones were brought into the sweat lodge and placed in the pit; we were told not to think about the heat; the stones were there to serve us. Then the door was closed. During the ceremony, there is no sense of time.

In life's experiences some things have a sacred side belonging to a higher order according to the Creator's intents and purposes. Tribes traditionally pattern community life according to nature's cycles and birds build their nests in a circle, neither of which depends on the clock.

As we continue to contemplate the world revealed by modern physics, it will become evident that it cannot be reduced to physical particles of matter like the lifeless "atoms" of the atomists in the 5th century BC. What, then, is the ultimate constituent of the Universe? So far, it appears to be the universal energy that pervades the Universe. Some have suggested that *mind* is the ultimate constituent. I'm not ready to accept the idea of mind *as a constituent* in the context of physics but I do not disagree with the notion that mind has always existed. I am not comfortable with any notion that places spiritual entities into a physics context. However, energy (or spirit) does manifest as matter and I am more comfortable with the notion that intelligence is inherent in nature and that mind (the Creator, the Great Mystery, the Great Spirit, the pervasive Power in the Universe) is the source of that intelligence and energy.

Where does this ultimate constituent comprising all physical reality reside? Is it within space-time or beyond? I tend to envision it as a layer, or substrate, of space-time; this concept is compatible with Paul LaViolette's notion of a primordial active and transmuting substance that acts like an ether (see "A new physics paradigm" in chapter 16). We are talking about something that is alive, active, transmuting, intelligent, and spiritual. After all, when we offer tobacco (or pollen or whatever substance we use and whatever our practice may be) or smudge with sage (or sweet grass), are we not linking two worlds (the manifested one and the invisible one) while also consciously acknowledging the Power associated with the experience and with the sacred elements?

We all dream. In dreams everything seems real. In my dreams pieces of furniture have argued with and fought each other. I often let myself fall off a high cliff and then I fly over a deep valley. I defy gravity by rising weightless from the floor or the ground. But when Australian Aborigines enter the Dreamtime, it is not a dream but a parallel universe. Wade Davis (*The Wayfinders: Why Ancient Wisdom Matters in the Modern World*) writes:

> As Aborigines track the Songlines and chant the stories of the first dawning, they become part of the Ancestors and enter Dreamtime, which is neither a dream nor a measure of the passage of time. It is the very realm of the ancestors, a parallel universe where the ordinary laws of time, space, and motion do not apply, where past, future, and present merge into one. It is a place Europeans can only approximate in sleep, and thus it became known to early English settlers as the Dreaming, or Dreamtime. But the term is misleading. A dream by Western definition is a state of consciousness divorced from the real world. Dreamtime, by contrast, is the real world, or at least one of two realities experienced in the daily lives of the Aborigines.[61]

Trustworthy friends of mine from different tribes have told incredible stories of events they have personally witnessed and experienced. They entered another dimension while fully awake, and in some cases, there were visible confirming signs. In one case, a friend of the one telling the story had died of an illness but later communicated to say he was doing well; when the conversation was over, there was an assuring hand-print on the arm of the one who told the story.

The use of indirect language to infer that scientific knowledge is authentic by comparison or that it is superior to Indigenous stories, considering also that the authors lack a correct understanding of Indigenous mythology, adds to public misconceptions. Hawking and Mlodinow (*The Grand Design*) refer to a legend of the Klamath Indian Tribes in Oregon, saying that it "faithfully matches every geologic detail of the event [the volcano that formed Crater Lake] but adds a bit of drama by portraying a human as the cause of the catastrophe" and that "the human capacity for guilt is such that people can always find ways to blame themselves."[62] It is truly commendable for them to acknowledge the legend's authenticity; however, their unfortunate allusion to guilt diminishes the significance of the legend.

I have heard and read many Indian stories of creation and emergence, including the Klamath legend, and each one was connected with a myth. The rain in the Klamath story was the result of the people's prayers and sacred songs but the notion of guilt is nowhere indicated in the story. These authors have interpreted a myth that is outside of their culture and apparently know very little about the purpose or character of Native myths. Asserting that most people don't ask philosophical questions about the nature of reality or why "there is something rather than nothing," they lament that "philosophy is dead because it has not kept up with modern developments in science, particularly physics."[63] While this may be true, it is science that is lagging behind in drawing conclusions that Indigenous peoples have long known and practiced.

As I state frequently in this book, Native peoples not only know the Rule of Law in the Universe (a chapter in the *Grand Design* book) but live in deference to it. The problem is that scientists isolate themselves from Indigenous realities and the general public reads their popular books but neither really pays attention to Indigenous knowledge. Besides, the philosophical questions posed by these authors are mostly academic, whereas Native peoples acquire knowledge that is necessary and practical—their knowledge *has* to work because often the alternative is death.

Spirituality should make a noticeable difference in the way we live in a secular and deeply divided society. The sense of sacredness is needed in a world dominated by secularism and consumerism where even many church-going people, by their habits and language, blend with the rest of society. Public statements and actions by Christians in politics address moral issues based on ideals but rarely render spiritual encouragement. The common practice is to look at virtually everything with dollar signs in the eyes. Based on what I've read about an era when sacredness was honored and spiritual power was more evident, I wonder if it will ever return.

The classical and quantum worlds represent distinct orders of reality. In the classical world, physical space serves as the background where all events occur, whereas quantum actions do not require a background and are not reducible to three-dimensional physical space. The background required by general relativity and the lack of it in quantum mechanics is one of the incompatibilities between the two theories which prohibits uniting them into a single quantum theory of gravity.

According to relativity theory, space is neither empty nor independent of time. Space-time is real and *bends* like a fabric in the presence of mass (mass is how we measure the amount of "substance" in a material object; it is what gives an object its inertia). Because general relativity theory is a *field* theory, one of its aspects is that it conceives of a Universe without rigid bodies—three-dimensional objects with definable boundaries. Instead, objects consist of locally intense concentrations of a force field from gravity. The gravitational field acts like a web in which particles are abstractions and there are no discontinuities.

In relativity, temporal and spatial properties vary with the frame of reference of the observer; these properties are now understood to be *relationships* rather than intrinsic features of separate objects in themselves. Every observer sees something different in other moving frames.

Reference Frames and Coordinates

Before we can examine relativity or consider objects in classical space , it will be necessary to introduce classical reference frames and coordinates. Matter, in turn, can exist as a gas, a liquid, or a solid.[46] A material object that retains its shape or form (for easier mathematical treatment), usually referred to as a "rigid body," contains three spatial dimensions. Any point in such a body can be located by measuring its distance from a fixed reference point in each of three mutually perpendicular directions, denoted by x, y, and z. This fixed reference point, called the "origin," is located at the point (x=0, y=0, z=0), denoted by the triplet (0,0,0). The point (2,8,3), for example, is located 2, 8, and 3 units from the origin in the x, y, and z directions, respectively. These three numbers are called *Cartesian coordinates* and the three perpendicular lines that intersect at (0,0,0) are called *coordinate axes*. A set of coordinate axes is also called a *frame of reference*. In relativity, a reference frame is treated like a rigid body;[47] *i.e.,* points in three-dimensional space are located relative to the origin (0,0,0) of the reference frame. The motion of a rigid body occurs relative to

---

[46] Matter can also exist in the form of plasma, which some view as a fourth form of matter; however, plasma is actually an electrified *gas* in which some of the electrons are free; the definition of "form" or "state" of matter is somewhat arbitrary..

[47] By thinking of the coordinate axes as three physical rods of infinite length, the reference frame can be visualized as a rigid body.

The Condor and the Eagle

another body, not relative to space. Perceiving reference frames as rigid bodies makes them consistent with the principle that motion occurs relative to other bodies, not relative to space.

There can be more than one reference frame, and frames can move relative to each other. Think of a bus in motion relative to the road with a reference frame located inside the bus and moving with the bus; a second, stationary frame is located on the roadside on the earth's surface and it is stationary. The frame on the vehicle is therefore moving at a certain speed relative to the stationary frame. Suppose that two people are riding in the bus and one throws a ball to the other in the direction of travel. If the ball moves at a speed of 30 miles per hour relative to the bus, and the bus is traveling at 60 miles per hour, the ball is then traveling at the speed of 90 miles an hour relative to the road.[48]

In the above example, we performed a *Galilean transformation* by adding the speed of the ball relative to the bus to the speed of the bus relative to the road. What if, instead of throwing a ball, someone shines a flashlight at a second person in the bus. Do we, as before, add the speed of light relative to the bus to the speed of the bus relative to the road? No, because *the speed of light is constant,* regardless of how fast the flashlight is moving relative to any observer. A Galilean transformation in this case, or any other situation in which the source of light moves, is not valid. In our everyday experience we can neglect the time it takes for light to travel between observed points, but it is an illusion. Also, because of relativity, a clock traveling with the bus ticks more slowly compared to the one on the ground, but the difference is not noticeable.

Because of relativity, an object traveling at any speed relative to a stationary frame of reference will shrink (contract)

---

[48] The ball is actually moving vertically and horizontally at the same time but the assumption here is that it is moving 30 mph in the horizontal direction.

along the direction of travel, and the amount of shrinkage (in length) will increase with speed. At speeds close to that of light, the amount is noticeable,[49] demonstrating that the geometry upon which Newtonian physics is based is not valid. Geometry assumes that the objects are static, that they do not change form or size.

Geometry contains the laws that govern the placement of rigid bodies mutually at rest relative to each other. Coordinate axes are also viewed as rigid bodies: three static rods of infinite length whose ends meet at the origin. In one of Euclid's axioms, the sum of the angles of a triangle is equal to 180 degrees. But this axiom holds only for triangles that are perfectly flat (two-dimensional); this geometry is *Euclidean*. However, for triangles on the surface of a sphere, the sum of the angles exceeds 180 degrees; this geometry is *non-Euclidean*.

Now consider a finite number of short rods placed end to end so that they form a circle, and let the circle spin. Since objects contract in the direction of motion relative to a fixed reference frame, the rods will contract lengthwise as they move along the circle. In Euclidean geometry the ratio of the circumference of a circle to its diameter is equal to the constant π (pi). Consequently, since the number of rods multiplied by the length of each rod is equal to the circumference of the circle, extra rods are needed for the circle to remain the same size. This implies that the ratio of circumference to diameter is no longer equal to π, providing another example of *non-Euclidean* geometry. We see that three-dimensional Euclidean space, the arena of Newtonian physics, is not valid in relativity physics. The above two examples demonstrate that Euclidean geometry does not hold for curved surfaces or bodies in motion at speeds

---

[49] The amount of shrinkage is computed by a mathematical formula known as the Lorentz contraction.

close to the speed of light. Space is non-Euclidean and not absolute.

Having introduced reference frames, we can state the principle of relativity as follows: *The laws of physics must have the same form in every frame of reference.*[50] That is, the equations describing the laws of physics have the same form in all reference frames. This principle holds for all relative motion. There are two cases of relative motion, *restricted* (or *special*), which refers to frames moving at a uniform relative speed, and *general*, in which one frame is accelerated relative to another frame. Frames that move at a uniform speed (constant and along a straight line) are called *inertial* frames. As I explain below, general relativity theory is actually a theory of gravity.

Note that physics as a science represents a unified system because of the premise that the laws of physics must hold in all reference frames; if the laws are true in one frame, they must be true in all frames. This is not only a "mathematical" concept but also a fact of nature.[51] It is true that the laws are expressed in mathematical form but mathematics is only a tool used to represent physical reality, not the reality itself. The laws of physics have demonstrated to me that they correlate with the empirical observations of Native peoples at least to the extent that facts can be perceived without mathematics, considering that both systems of knowledge attempt to describe the same reality.

Special Relativity

Einstein was troubled for years about the incompatibilities between Newton's laws, which assume the absolute nature of

---

[50] The principle of relativity and the theories (special and general) of relativity are not the same thing. The two theories will be treated in the next two sections.

[51] Note the subtle difference between the laws of physics, which are part of nature, and physics as a science, which is a human discipline.

space and time, and Maxwell's equations, which treat light as a constant in all reference frames; they could not both be right. One day in early May of 1905, he went to visit his good friend Michele Besso, an engineer, who also worked at the patent office, to explain the dilemma and told him that he was going to give up trying to solve the dilemma. But as they discussed it, he suddenly recognized the key to the problem: *Time cannot be defined as absolute.* Two events that appear to be simultaneous to one observer will not appear to be simultaneous to another observer if the two observers are moving relative to each other. Five weeks later, he submitted the paper for publication.

Einstein explained his theory using an image of the station in Bern where trains moved past the rows of clocks that were synchronized with the clock on top of the famous tower. He said that judgments in which time plays a role are always judgments of simultaneous events. He further explained that when he says a train arrives at 7:00 o'clock, he means that the small hand of his watch pointing to 7 and the arrival of the train are simultaneous events when, in reality, observers who are moving rapidly relative to one another have a different view of the two distant events; their clocks do not agree.[64]

Imagine, with Einstein, that you are racing away from a clock on the Bern tower at the speed of light. What would you see? The clock would appear frozen in time (it would not tick), since light *could not catch up* to you, but your own watch would tick normally. If you moved any faster, the clock would disappear. Likewise, time lapses at different rates throughout the universe, depending on how fast you move relative to other places. If we imagined clocks (which had been synchronized) scattered at different points in space, each one would tick at a different rate and show a different time relative to an observer who could see all of them at once.

The special theory of relativity rests on the following two postulates:

1. The speed of light in a vacuum is the same for all observers, regardless of their relative motion or the motion of the source of the light.

2. The laws of physics are the same for all observers in uniform motion relative to one another.

Galileo stated that all uniform motion is relative and that there is no special (privileged) reference frame where the laws might be simpler or from which the laws in other reference frames are derived. Einstein extended this concept to account for the constant speed of light and postulated that it holds for all the laws of physics, including the laws of mechanics and electrodynamics, whatever those laws may be.

The principle of relativity on which the special theory rests is *restricted* to uniform relative motion between inertial reference frames; it was stated by Einstein as follows:

> If a system of coordinates K is chosen so that, in relation to it, physical laws hold good in their simplest form, the same laws hold good in relation to any other system of coordinates K′ moving in uniform translation relatively to K.[65]

The above principle can also be stated as follows: *The laws of physics are the same for all inertial frames of reference.* "Uniform translation" means motion in a straight line at constant speed. The principle defines an *inertial* frame of reference; *i.e.*, that all inertial frames are in a state of constant, rectilinear motion with respect to one another. The *special* theory of relativity limits motion to inertial reference frames (the second postulate above), whereas the *general* theory, discussed in the next section, removes this restriction by assuming that one reference frame is *accelerating* relative to another frame. At that time, it will be explained how acceleration and gravity are related. Both theories depend on the constancy of the speed of light.[52]

In his 1905 paper on the special theory of relativity, Einstein postulated that the velocity (speed) of light in empty space (a vacuum) is a universal constant (the first postulate above), which was previously observed in the Michelson-Morley experiment. When Einstein discovered the laws of special and general relativity, he eliminated the notion of absolute space and time and introduced in its place a new, *relational* approach in physics. As a consequence of light's constant speed (and of relativity theory, which rests on it), several counterintuitive effects are known to occur: length contraction, time dilation and the relativity of simultaneity:[53]

- Length contraction: The length of a moving object contracts in the direction of motion when measured by an observer at rest. This implies that space is not absolute. Another consequence is that the mass of an object increases with speed, resulting in the concepts of "rest mass" and "relativistic mass"; thus, the total mass of the object is the sum of these two; however, it is impossible for any particle possessing rest mass to be accelerated to the speed of light.

- Time dilation: The time lapse between two events is not invariant from one observer to another but dependent on the relative speeds of the observers' reference frames. A clock traveling with an observer runs slower compared to a clock stationed with an observer at rest (assuming, of course, that the two clocks are synchronized). Thus, time is not absolute; time lapse depends on relative motion. As an observer moves away from a light source, the light waves become more

---

[52] A common misconception is that "constant speed of light" means that light has not changed speed throughout history. But in the special theory, it means that it is the same relative to inertial frames, not relative to historical time.

[53] These statements are not precise or complete. Although the Lorentz transformations are easy to derive, they are omitted to avoid discussing the mathematics.

spread out (and the light will shift toward red in color). But since the speed of light is constant, time itself must dilate.

- Relativity of simultaneity: It is impossible to say with absolute certainty that two events occur at the same time if the events occur in different spaces. The perception of when each takes place depends on the observer. Only if the two events are both local in one reference frame will they be simultaneous in the other frame.

The time dilation phenomenon has been confirmed experimentally using atomic clocks on airplanes, fast-traveling particles, and in other ways. In the famous "paradox of the twins," someone goes on a space voyage, traveling near the speed of light; upon returning, the twin who stayed at home (on Earth) is discovered to be much older. This so-called paradox is only *apparent* because what is described is factual, not a paradox.

In relativity, objects always move relative to other objects. Space and time are mutually dependent and space-time is flexible, like a fabric. A point in space-time is located by specifying four dimensions: three spatial dimensions and one time dimension. Temporal and spatial properties vary with the frame of reference of the observer; these properties are understood to be *relationships* rather than intrinsic features of separate objects in themselves because every observer sees something different due to the relativity of motion.

The equations that transform space and time coordinates between two inertial frames of reference are easy to derive but not included here; these are the Lorentz transformations. At sufficiently low speeds, the equations approximate Newton's laws, and in the classical limit[54] they are the same. Particles

---

[54] The "classical limit" is approximately where the relativistic formula merges smoothly with the Newtonian formula when a value, such as speed, is small

without mass, such as photons, can move only at the speed of light, while particles with rest mass can never reach the speed of light. Light travels slower in a medium (such as water) than in a vacuum (empty space).

General Relativity

The special theory of relativity assumes that the relative speed between observers in all frames of reference is uniform; *i.e.*, they are inertial frames. The next problem that Einstein undertook was to develop the general theory of relativity without this restriction; it was the obvious next step. He worked arduously for eleven years to develop the theory for accelerating frames of reference, publishing it in 1916. Newton had derived his law of gravitation experimentally, whereas Einstein derived his equations theoretically. This is an important difference: the theoretical approach made precise assumptions (the constancy of the speed of light, for example), giving it the ability to make predictions; the experimental approach was based on approximate measurements.

To develop the theory of relativity for accelerated relative motion, he had to resolve the following issues: a) a conflict between Newtonian gravity and the special theory of relativity; b) the equivalence of two kinds of mass; and c) the difficult mathematics required to solve the complex problem of geometry. The theory finally emerged and was later validated experimentally. Let's look at what Einstein did.

First, he noticed a serious conflict between Newton's description of gravity and the special theory of relativity. It is possible not to notice this conflict and believe that we understand gravity because we don't recognize a hidden assumption. Students of elementary physics are very familiar

---

enough to be neglected. This is known more generally in physics as the *correspondence principle.*

with the inverse square law. According to Newtonian attraction, the gravitational force between two bodies is directly proportional to the product of their masses and inversely proportional to the square of the distance between them. For example, doubling the distance between the two bodies *weakens* the gravitational force by a factor of four; reducing the distance by one-half *strengthens* the force by a factor of four.

It is not difficult to understand why Newtonian attraction is problematic without delving into theory: it implies that the force of attraction between any two bodies acts *instantaneously*, no matter how vast the distance between them may be. This idea even troubled some of Newton's contemporaries. The inverse square character of Newtonian gravitation is only a good approximation at sufficiently short distances; nonetheless, it was deduced from experimental results rather than crucial theoretical considerations.

Hypothetically speaking, if a massive star or planet suddenly collapsed (or mysteriously disappeared), how soon would the effect be noticed? According to Newtonian attraction, the effect would be immediate, which violates the principle that nothing, including gravity, can travel faster than light. General relativity theory resolves this problem by showing that gravity travels through space-time at the (finite) speed of light.

To address the second problem, we recall that general relativity theory removes the restriction of inertial frames of reference. But how does this involve gravity? The answer has to do with a unity that exists in nature, the key concept that allowed Einstein to take the initial step in deriving the general theory. Let me explain:

There are two ways to measure mass. According to Newton's second law, a material body (object) experiences an acceleration a when a force F, and no other forces including gravity, is applied to it;[55] *i.e.,* force equals mass times acceleration, or $F = ma$, where *m* is the body's mass. This mass is called *inertial* because the applied force works against the body's inertia.

The other kind of mass, called the *gravitational* mass, pertains to the weight *W* of a body; I will denote this mass by *M* (capitalized). The weight is the gravitational mass *M* multiplied by the intensity of the gravitational field *g*; *i.e.,* $W=Mg$. It is only when the two masses are equal that the two force are equal. One way to think about this is to imagine yourself in free-fall; you are weightless because there is no force under you and the only force is gravity (assuming that there is no air resistance), which is equivalent to your weight. This same force (gravity) is overcoming your body's inertia. The conclusion is that, since the weight *W* and the force *F* that overcomes inertia are the same, the inertial and gravitational masses are equal:

$$F = ma; \ W = Mg$$
$$\text{If } F = W \text{ then:}$$
$$ma = Mg$$
$$a = (M/m)g$$

The condition necessary for the acceleration, *a*, to be the same for *all* bodies in free-fall regardless of their composition (what they are made of) is that the ratio (M/m) must be the same for all bodies. For the same body, the gravitational mass and the inertial mass are equal ($M = m$). The equivalence between gravitational and inertial mass is called the *equivalence principle*. Air resistance and other possible contingencies, such as gravitational influences from other bodies, have an effect on a

---

[55] It the force *F* is applied horizontally, the horizontal motion will not be affected by gravity.

falling body but do not invalidate the equivalence principle. Einstein stated the principle the following way: The gravitational "force" [weight] experienced while standing on the earth is the same as the "pseudo-force" experienced by an observer in a non-inertial (accelerated) frame of reference, or more simply: *Every case involving inertia can be replaced by an equivalent one involving gravity.*

The equivalence principle has been stated in different ways, for example: (a) "It is impossible for someone standing inside a stationary box (like on an elevator that isn't moving) who experiences a force on his feet (due to gravity) to know, from the force alone, whether that force is the result of a gravitational field (the box is standing on the earth's surface) or an acceleration (the box is being propelled by a rocket)." Also: (b) "Standing still on the earth's surface feels the same as being inside a spaceship (away from any sources of gravity) that is being accelerated by its engines."

Einstein developed the theory of general relativity by dispensing with inertia and using gravity as the basis instead. Inertia also seemed to suffer a weakness due to circular reasoning. Newton's first law of motion, also called the law of inertia, states: "Every body continues in its state of rest, or of uniform motion in a straight line, unless it is compelled to change that state by forces impressed upon it." In Einstein's view, Newton's first law used circular reasoning: a mass doesn't accelerate if it is sufficiently far from other bodies, and we know it is far from other bodies because it doesn't accelerate. Are there large regions of space in the Universe that are free from the influence of gravity?

The third issue that Einstein had to resolve was the difficult mathematics. Recall that, according to the special theory, a body shrinks along the direction of motion and that Euclidean geometry is no longer valid. Space-time geometry is, of course, noticeably affected only at speeds approaching the speed of light. Since a rotating body experiences shrinkage along its outer edges, the body's shape becomes distorted, which changes the geometry and in turn the gravitational field surrounding the body. You can see that mass has an effect on geometry and geometry has an effect on gravity. *How* this happens is complicated; general relativity theory explains this interaction, as we shall see.

Einstein's mathematical development of surface geometry reached an impasse because the geometry required was beyond his knowledge. The problem involved multi-dimensional surfaces (surfaces of higher dimension) as well as *irregular* surfaces. Its resolution was necessary because it is unrealistic to assume that space-time geometry will change in predictable ways as the gravitational field changes; Einstein's theory had to describe the interaction between gravity and space-time changes at any point in space-time on the randomly irregular surface. Fortunately, the mathematics he needed had already been developed by a German mathematician by the name of Georg Bernhard Riemann in the 1850s, long before the development of the theory of relativity. Riemann is, in fact, the founder of Riemannian geometry. This was the final step in the process: to advance from basic Gaussian geometry theory, which treats curves and surfaces, to the advanced theory of Riemannian geometry.

As already stated, space and time can no longer be conceived as independent entities; they are mutually dependent and both are altered in the presence of mass. *Time, space, and gravity cannot exist separate from matter.* By applying general relativity theory to the earth's motion around the sun, for example, we can visualize the earth as a golf ball falling into a depression on a flexible sheet created by a bowling ball resting on it, where the sheet represents deformed space-time and the bowling ball represents the sun. The picture depicted here is that space-time acts like a flexible fabric. In other words, according to Newtonian gravitation, the earth is attracted toward the sun and revolves around it, whereas in general relativity theory the earth follows a space-time path created by a large mass. The presence of mass dynamically changes the space-time geometry around it.

In Newtonian physics, matter (mass) produces the gravitational force, but in general relativity, *mass is not the only source of gravity*. Relativity links mass with energy, and energy with momentum. The energy of any physical system derives from all of its properties that are associated with energy or mass (since these are equivalent); these properties act as sources of gravity and they include a body's temperature as well as the binding energies of nuclei and molecules.

Einstein's equation (actually a system of ten equations) relates space-time geometry to the properties of matter. As already stated, Riemannian geometry was necessary; the geometry is complicated in two ways: the irregular curvature of the surface and the multiple dimensions of the irregular surface. The Einstein equation of general relativity may be written as follows:

$$\mathbf{G} = (8\pi G/c^4)\mathbf{T}$$

In the above equation, $\mathbf{G}$ (in boldface on the left side) is a geometric quantity, called the *Einstein tensor*, which measures curvature; $\mathbf{T}$ measures matter content (energy, stress, momentum). They are related through the constant shown in parentheses containing the Newtonian gravitational constant G and the speed of light c. $\mathbf{G}$ and $\mathbf{T}$ are *tensors*. If you are familiar with vectors (quantities that have magnitude and direction), a vector is also a tensor of order one; **G** and **T** are tensors of order higher than one.

The geometry of space-time uses a key quantity called a *metric* that gives distance and angle (curvature) values at any point in space-time. The distance, d, between two points on a two-dimensional (Cartesian) surface is given by the length of the hypotenuse d of a right triangle, given by $d^2 = x^2 + y^2$, where x and y are the lengths of the other two sides. For a more general (irregular) surface, we can write $d^2 = (g_{11})x^2 + (g_{12})xy + (g_{22})y^2$; you can see that, for a plane surface, $g_{11} = 1$, $g_{12} = 0$, and $g_{22} = 1$; these g scalar quantities are called *metrics*.

A crucial condition that must be met, and which expresses the constant speed of light between two frames of reference, one of which is accelerating relative to the other, is that the distance d between any two points on the surface *must remain constant when expressed in the coordinates of either reference frame.* Space-time has four dimensions, requiring three spatial coordinates and one time coordinate. But the mathematical treatment is simplified if the fourth coordinate (time) is expressed in terms of distance. This distance is expressed as ct, where c is the speed of light and t is the time it takes for light to travel the distance d. This conversion gives all coordinates the same units.

The foregoing discussion explains how the first level of geometrical complexity that Einstein encountered—that of Gaussian geometry, which uses *curvilinear* coordinates instead of the simpler rectangular coordinates—was resolved. But it didn't solve the problem of multidimensional surfaces, requiring Einstein to apply Riemannian geometry. All of the metric information (geometry and gravity) is encoded in the Einstein *metric* tensor, **G**, for each point in space-time. The metric tells exactly how space-time is curved at every point.

The late theoretical physicist, John A. Wheeler, summarized the Einstein equation using the following familiar expression:

"Space-time tells matter how to move; matter tells space-time how to curve." Space-time is represented by **G** and matter is represented by **T**; therefore, **G** influences **T**, telling matter how to move, and **T** influences **G**, telling space-time how to curve.

The equations of general theory predicted an expanding Universe, a surprising result that Einstein did not expect, impelling him to add an arbitrary term, called the "cosmological constant," to the equation in conformity with his belief in a static Universe. It turns out, however, that he tried to fix something that wasn't broken; it was later discovered that the Universe *actually is expanding* and the constant is appropriate after all. This mistake, he said, was his "greatest blunder." Furthermore, recent research has revealed not only that the Universe is expanding but that the expansion rate is also greatly accelerating—a phenomenon that represents a new unsolved mystery in physics: *dark energy*, adding to the mystery concerning *dark matter*.

But an expanding Universe implies that space-time is also expanding, further suggesting that the physical world was much smaller in the past. Everything—planets and smaller bodies, stars, galaxies, clusters of galaxies—were closer together in the past, all linked together by the gravitational field acting like filaments of energy. This does not mean that celestial bodies are moving out into space; instead, space-time is itself stretching and moving the bodies with it.

Einstein passed to the spirit world without realizing his lonely quest for a unified field theory that would tie together electricity and magnetism, gravity, and quantum mechanics, giving a more complete explanation of the Universe. He sought a theory that would mathematically link the gravitational field and the electromagnetic field as components of one uniform field. Other physicists are employing different approaches and have so far been unable to unify quantum mechanics (which describes three of the four fundamental interactions, or forces, of nature) with Einstein's general theory of relativity (which describes the fourth force, gravity); this unified theory has been called the *quantum theory of gravity*, or *quantum gravity*. It would unify all of the fundamental forces of nature at subatomic and cosmological scales into a single mathematical framework. One of the reasons the two theories are incompatible is that relativity requires a space-time background where all actions occur while quantum theory does not. A field theory involving mass-less particles called *gravitons* could do this.

Energy and Motion

The well-known Einstein equation, $E=mc^2$, where m is a body's mass and c is the speed of light, expresses the equivalence of matter and energy and reveals the connection between mass (matter) and the speed of light. This equivalence is the consequence of the special theory of relativity. Einstein derived this equation using a simple method without referring to mass by envisioning a box containing radiant electromagnetic energy; the box is accelerated so that radiation pressure is exerted against the rear wall of the box. He showed that this radiant energy had the *same effect* as a real mass resisting acceleration. This proved that inertia resulted from the internal movement of radiation rather than mass, showing that energy and mass are truly equivalent, for it was light, not matter, that produced inertia. This emphasizes once again that energy is connected with movement. Inertial mass and gravitational mass are only one aspect of movement; energy is another form. The transformation of matter into energy is a change from one form of movement into another form.

Possibly the most awesome aspect about the equivalence of mass and energy is the incredible amount of reserve energy contained in a body at rest. Objects are in a state of violent movement that is hidden from our eyes. We can immediately see from the equation $E=mc^2$ that this is a very large amount of energy, because the speed of light, c, is a very large number: 3×10^8 (300 million) meters per second, or 186,000 miles per second. Squaring this quantity and multiplying it by the mass m gives a very large number, even when m is small.

The fact that the internal movement occurring within a piece of matter contributes to its mass has been verified experimentally in several ways. When the positron (a negative electron, or antiparticle) was discovered, it was found that when it collided with an electron, the two annihilated each other, so that only energy and no particles resulted from the collision. Since that discovery, other antiparticles have been discovered, and antiparticles have also been created from energy. These experiments and discoveries validate the formula $E=mc^2$ and verify the tremendous amount of energy hidden inside matter. The fusion processes taking place in the sun and stars exemplify energy/matter transformations occurring spontaneously in nature. In fusion, heavier atoms are formed; in fission, atoms split into lighter elements.

From the foregoing it should be easier to understand that transformations of matter into energy produce a very large amount of kinetic energy, which in turn produces a large amount of heat. Einstein spoke of the great threat of evil this large amount of reserve energy implies.

Bohm showed how the rest energy of a body is derived from the total kinetic energy of its constituent particles.[66] He also cited experimental evidence to show that the internal state of movement of these particles contributes to the total (rest) mass of the body and therefore to the rest energy $E = mc^2$. "So there has been conclusive experimental proof that either a part or the whole of the 'rest energy' of a body can be transformed into other forms of energy, and that the inverse process of transforming other forms of energy into rest energy is also possible." [67]

The connection between movement and Native languages is inescapable. Two examples that come to mind are the Maya and Pueblo tribal peoples. Author Joseph Rael from Picuris Pueblo, who grew up speaking the Tiwa language since the age of seven, writes: "Then, life was created by *Wah-mah-chi*, which is the Tiwa name for God, and means breath, matter, and movement. The breath is the inspiration in matter that brings all concreteness, or form, into existence via movement. Really only one thing exists, and that is the breath of God in a state of movement creating the vibration of matter."[68]

According to Mayan holy man Hunbatz Men, Hunab K'u is the unique Giver of Movement and Measure and represents Absolute Being. Speaking of his Mayan ancestors, he writes: "Their concept of Absolute Being was defined as measure and movement—measure of the soul and movement of the energy which is spirit."[69]

According to relativity theory, the mass of an object increases with speed. If energy and movement are related, as stated above, how can a body at rest have any energy, since it is stationary? A material body of any size consists of smaller particles. The constituent particles within a body are in a state of violent movement.[70] As we look deeper into the constituent molecules, we find that they consist of even smaller particles— atoms, which consist of nuclei (protons and neutrons) and electrons. Digging still further in search of smaller particles will lead to quarks and the strings of M-theory. According to particle theory, even protons, neutrons, and mesons are not elementary particles but composites of quarks.

Chapter
8

ASPECTS OF SCIENCE

I s there only one science?

Several years ago, while serving as dean of science and mathematics at a tribal college, I received an official visit from a representative of a U.S. government office. As we were becoming acquainted and conversing over a variety of subjects, I happened to mention "Native science" with reference to the methods and philosophy that Indigenous peoples apply to their way of life. This term apparently annoyed or surprised him and, before I had the opportunity to elaborate, he said, in a dispassionate but authoritative manner: "There is only one science."

But he was wrong. He was also arrogant though probably unaware of it. He did not ask what I meant by "Native science"; instead, he assumed that his position was superior. The widespread notion of one science undoubtedly stems, at least in part, from lack of contact between Western and non-Western peoples in the history of Western science. Yet there are actually many subsidiary sciences, such as astronomy, medicine, horticulture, and the Indigenous science behind the numerous varieties of potato crops that are genetically suited to different altitudes in the Andes.

The prevailing view is that "science" is an all-inclusive term containing many branches of specialized disciplines, such as physics, chemistry, and biology. Some say it is "a body of knowledge," while others say it is the *pursuit* of knowledge or the pursuit of truth; still others refer to the dictionary. All are correct but when a scientist, like this accomplished professional in my office who possesses a doctorate in chemistry and occupies a university faculty position, asserts that there is only one science, the reference is most likely to the empirical methods used for acquiring knowledge. A longer definition of science, which reflects Western epistemology, might be something like the following: science is the gathering, testing, and systematic organization of empirical data using refined procedures (*i.e.*, the scientific method) resulting in the vast body of knowledge that humans have amassed over time.

Let us look briefly at the history of science.[56] We know that the term "natural philosophy" (the study of nature) preceded "science." Some time in the past, probably before the nineteenth century, science was not a separate discipline like it is today and there were no people called "scientists." Originally, "science" simply meant knowledge, which is the literal translation of *scientia*, the Latin word for knowledge. The investigation of nature by early Greeks was called "natural philosophy," a name that endured well into the Middle Ages. Their mode of inquiry was *qualitative—why* do things happen? Natural philosophy, also called "Greek science" in the form of Aristotle's works, became firmly established in undergraduate education by the middle of the fourteenth century, when a new master's degree was introduced at Paris in 1341 requiring that the "system of Aristotle" be taught.[71]

---

[56] The history of Western science will be discussed in much more detail in a subsequent chapter.

The Condor and the Eagle

Modern science began in the 16th and 17th centuries when a new, *quantitative* mode of inquiry came into usage, meaning that investigators of nature were speculating on *how* things happen; they became obsessed with empiricism and objectivity, detailed facts, the "experimental method," and the application of mathematics. The works of Aristotle, Plato, and the genius of other earlier Greeks, were still available. But at the end of the Middle Ages, atomism returned to science.

Among the key figures during the rise of modern science were Copernicus, Galileo, and Newton. Science progressed rapidly due to inventions like the microscope, the telescope, the mercury barometer, the graphite pencil, and the slide rule. Much of the success of Western science is due to mechanism & quantification.

Key figures from the 18th and 19th centuries were Newton, Lagrange, and Maxwell who published books during this period. This time in history also saw the rise of materialistic mechanism and reductionism, the Enlightenment, and the Industrial Revolution—undoubtedly an age of triumph. But materialism and the notion that everything can be explained by means of materialistic particles hardens into dogma.

As we have seen, during the 20th century nature began to reveal some of her secrets and, in my opinion, is now beckoning physicists to shift toward holism in science. Reductionism can no longer be defended. Indigenous peoples have always contemplated the cosmos as an interconnected and undivided whole, as modern physics shows. Science nowadays has become the virtual arbiter of truth but I believe that only with the heart will we see and experience deeper realities.

Nature

Elders tell stories during tribal community gatherings, many of which my wife and I have attended, particularly on the Lummi

Indian reservation and the root feasts of the Plateau tribes in the Pacific Northwest, and we have heard the elders say "don't try to change nature."

There is little respect for nature. If we are taught to ask permission when taking fruit from a tree or when coming upon a place that was once occupied by a people from a different era, and we see the structures still standing, we learn that not everything is ours until it is given. I suggest we do the same when interrogating nature; we should do it respectfully and consider the relationship. We are then likely to refrain from the kinds of activity that are destructive and invasive. Guidance from Spirit can bring light to our minds to make the right choices.

But how do we do this? The late historian Joe Sando (*Paa Péh*) from Jemez Pueblo describes a time when the people lived in loving relationship with the land that sustained them:

> The secret of their existence was simple: They came face to face with nature but did not exploit her. They became a part of the ecological balance instead of abusing and destroying it, accepting the terms of their environment and becoming a part of the land, utilizing what the land had to offer without changing it.[72]

They were not "scientists" but they related to the land and became a part of it. Relating to nature is not science in the usual sense but interacting with nature is still a part of what scientists do. Birds learn to fly, eagles make their nests and raise their young, and a wolf pack uses group strategy to bring down a large animal; they know by instinct or they learn by communicating with their own kind and they do it without our help.

Guiding a child to closely observe how the ants are always working and bringing food home from far away without getting lost; leading them to observe how birds make their nests and the

The Condor and the Eagle

parents feed their young until they fly out of the nest are experiences that will help children know that nature is endowed with intelligence, consciousness, and spirit. There is no need to prove that there is a Designer, which makes it a cause-and-effect argument for science. But it is not science in the conventional sense of learning about nature; it is learning *from* nature. When we teach science, we convey with the right attitude what we have learned from nature, our trustworthy teacher. Our perspective (what we think about what we teach) and our approach (how we teach) will determine whether students (and children) learn to relate to nature or become alienated from her.

Knowledge

Native peoples have always known that learning involves the whole being: intellectual (mental), physical, emotional, and spiritual. Otherwise, the pursuit of knowledge is deficient in the process of learning, resulting in an *incomplete* picture of the world. It is important to recognize that Western science presents an incomplete view of nature and the world by omitting certain sense experiences, such as spiritual ones, that cannot be ascertained or repeated through measurement; insofar as American public education advocates this view, it is incomplete.

Science is not wrong because it focuses exclusively on the empirical; this is a matter of choice. But it *is* wrong for insisting that only empirical sense experiences are valid; that what cannot be measured does not constitute "science" insofar as it means knowledge or truth; that there is only one science: Western science. Such dogma also leads to public ridicule of non-Western cultures and it abrogates the responsibility to discern.

Philosophy

Schools of philosophy have also influenced the meaning of science even in the "hard" sciences like physics. Especially in quantum theory, which raises questions about the nature of reality, the influence of philosophical thought is evident among physicists who contributed significantly to its development. Among these were Neils Bohr, Albert Einstein, and Werner Heisenberg, who did not concur on meaning because of philosophical differences.

The dualism in Western culture is problematic when it comes to understanding mysticism or spirituality. Ideas that claim the existence of spirit are viewed as superstitious or religious because they invoke the supernatural (which means "outside of nature") and explanations based on the supernatural are considered "unscientific." On the question of the existence of a higher Power, the dualistic argument usually goes like this: Since the Universe by definition consists of everything in existence, which also implies that nothing can escape or enter, then nothing on the outside (God, for example) can explain (or control, influence, create, etc.) the things that are inside, because "outside" by definition does not exist. Thus it is common to reject the mysterious, deny the spirit, or demand empirical proof. The problem, I believe, is that all of these assumptions deny that spirit is embedded in nature or that nature is imbued with spirit. In a spiritual/physical Universe, there is no need to go outside, for all things are manifestations of spirit. Nor is space-time the lowest level of reality; there are values which exist or originate outside space-time.

On the Origin of Life

When then-Senator Barak Obama and Senator John McCain were campaigning for the office of President of the United States in 2008, each was asked in front of a large church audience, "When does human life begin?" Mr. McCain said it begins at conception and Mr. Obama said "it's above my pay grade." The "pay grade" response seems the more honest reply: I took it to mean "neither I nor science knows the answer." Indeed, that life begins at conception may agree with personal belief or biblical interpretation but it cannot be verified or tested by science.

Neither response was correct if we consider the possibility, as I do, that life has always existed. If so, then life exists before conception and it existed before science. All things are embedded with spirit, for Creator breathed into everything. Every object is the manifestation of its own spirit, just as genes give instructions to an organism; the Creation is also a manifestation of the invisible, spiritual Universe. Evolution, if it is true, neither creates life nor originated life in the Universe. This view is compatible with the Bible, conservatives, liberals, progressives, and even evolutionists. By accepting it as a common basis, many conflicts might be avoided.

Chapter
9

SCIENCE AND CULTURE

From the perspective of many physicists, the equations of modern physics may come closer to describing physical reality than ordinary language. Some philosophers of science, on the other hand, assume instead that science is done within the context of a comprehensive worldview forged from culture and primarily constructed in ordinary linguistically-based language—an assumption that could threaten the authority of scientific truths.[73] I mention these views to bring attention to the link between culture and the authority of scientific knowledge.

Regarding the first of these two assumptions, the equations of mathematical physics do have explanatory and predictive power but the view, if taken to an extreme, may be taken to imply that in *some* cases physical reality can only be described in mathematical form. This would mean that not even Native languages can describe some aspects of nature. It is doubtful that this is the correct interpretation; however, much depends on what is meant by "ordinary language."

There is no doubt that culture has an influence on science but to say that science is conducted within the context of a worldview requires clarification, particularly what is meant by "science"; Indigenous societies who have lived independently of

The Condor and the Eagle

the Western World's science have their own science that is learned and practiced completely within their culture. The main point about this whole issue is that the assumptions of the West often seem valid until Native peoples are taken into account, which typically doesn't happen, requiring those assumptions to be re-considered.

Philosophical Aspects of Cultural Difference

Edwin J. Nichols, a behavioral psychologist with Nichols and Associates, uses a unique paradigm and pedagogy that he developed in the 1970s called "The Philosophical Aspects of Cultural Difference."[74] It is a deep philosophical perspective on the rationale that led to the differences that exist among cultures and, in particular, helps to explain the causes of historical conflicts between Natives and Europeans. I include it because his approach is found nowhere else.

He examines four groups of people that he says share a common worldview: (1) European or Euro-American; (2) African, African American, Latino/a, Arab; (3) Asian, Asian American, Polynesian; and (4) Native American. Each worldview is associated with a unique history. He examines each group on the basis of three philosophical aspects, or essential factors, of cultural difference: *axiology* (the study of values), *epistemology* (how do you know knowledge), and *logic* (how do you reason). Nichols is an effective communicator; he employs a conversational approach and encourages audience interaction.

On the axiology of Europeans (and European Americans), according to the model, their highest value is in a member-to-object relationship; *i.e.*, the important factor, or the highest value, is in the "object," something external. For example, *power*, *control*, and *authority* are paradigmatic objects in the European worldview. Thus, the European tacitly always seeks the object. When a member of this group loses the object—the job, the house, an investment—it can have a traumatic effect. As partial evidence, he cites the extraordinary number of deaths that occurred in 1970 as a likely consequence of the stock crash of that year.

He uses this analogy: If I have in my possession all of the objects and you need one from me, a *hierarchy* can develop in which I, the owner, have *control* over lower members of the class, of which you are one. You become subservient to me, as I can then give you orders. In this structure, rank means power and authority.

On the axiology of Native Americans, according to this model, their highest value is "oneness with the Great Spirit." According to the traditional metaphysics of Indigenous culture(s), all things are related and interconnected: the two-legged, the four-legged, the winged, things with fins, things that crawl, etc. Each has a member-to-Great Spirit relationship in an ecological balance. So, when they came in contact with Europeans, people with a member-to-object axiology, the two groups clashed. The Indian needed buffalo to survive and the European wanted the *object* for its hide, which is a *part* of the animal but still an object. To the Europeans, the hide was valuable in terms of money and profit—objects of value. The killing of the buffalo, the cutting down of trees, and other destructive acts destroyed nature's balance as well as the essential elements of survival for the Indian. But to the European, they were just objects.

For the European, science is done from an objective (object to object) perspective, data are collected and analyzed and the results are validated. In most cases, profit is involved in the process. But for the Indian, it is a personal, respectful, and balanced relationship.

The Condor and the Eagle

Culture does influence science. Other value differences that contributed to the Indians' historical trauma could be cited: views about the land (considered a commodity, wasted if idle), time (it is linear, it can be "wasted"), the cosmos (dead, passive, and inert), the earth (contains resources for development)—all are objects that the European considers separately, whereas to the Indian they are a part of a unified world. Moving to a reservation meant loss of freedom, the inability to live off the land, and confinement to a small and incomplete world. Sociological and economic problems were inevitable. On the other hand, reservations also had positive effects stemming from the continuation of community life.

On the epistemology aspect, which relates to knowledge and therefore to learning and education, Nichols tells a story: If you were a young Greek long ago, you would be sent to study at Memphis in Africa where an education takes twenty years to complete. If you left at age 15, you would return at age 35. But the life expectancy in Greece is 27, so what do you do? You reduce what you have to learn to the lowest common denominator; this means limiting *how* you learn by deciding on the method of counting and measuring; whatever cannot be counted or measured does not qualify as learning; it is not knowable.

If in Africa you need mind, body, and spirit to learn, says Nichols, what do you eliminate in order to reduce learning to counting and measuring? Can you measure the mind (intellect)? Yes. Can you measure the body (physical)? Yes. Can you measure the spirit? No; eliminate it and give priority to *quantitative* knowledge.

But a dilemma still remains: whereas knowledge is stored in the mind, in just a short time the body will die (due to the brief life expectancy of 27 years). The solution is to separate mind from body, which leads to *dualism*.

Tacit Infrastructure

Every summer from 2002 until 2009, I participated with a group of quantum physicists, linguists, and American Indian scholars who convened every year at the Language of Spirit Conference in Albuquerque, New Mexico to discuss the underlying principles of the Universe. Sponsored by SEED Graduate Institute, the sessions were conducted in the tradition of the late physicist David Bohm, a colleague of Albert Einstein, who authored several books on quantum physics and philosophy and also became known for his penetrating approach to communication referred to as "dialogue."

It was there that I was introduced to the useful notion of suspending our individual points of view that make us rigid in conversation and function as a major barrier to effective dialogue because they hinder the free flow of ideas such that we don't give full attention to the meaning that others are trying to convey. A "tacit infrastructure"[57] operates within each person and in society, consisting of ideas, cultural values, and perspectives that cause us to act without thinking about the deeply rooted assumptions upon which they are based. It is important to temporarily suspend that part of our infrastructure—not all of it, of course!—that wants to debate, insult, or win arguments.

The decision to suspend is intentional, a useful discipline that can train us to change the way our mind works for everyone's benefit. Depending on the situation, such as often occurs in Indian groups and communities, the goal may be to arrive at a consensus or to achieve a common consciousness in which there are no winners of losers. This is vastly different from the divisiveness that dominates politics.

---

[57] This term was used by physicists David Bohm and David Peat in their book, *Science, Order, and Creativity.* I've take the liberty to paraphrase and

Origin of Western Science

Western science stems from a Western epistemology. However, there are scientists from the Western tradition, including Bohm, now recognize the constraints imposed by the scientific method and are calling for a new paradigm in science. In contrast to the assumption in Western science that "knowledge exists only if it can be measured," Native peoples embrace the whole of human experience rather than limiting themselves to what can be gathered and interpreted through the physical senses; *i.e.,* through "sense experiences." An essential element of Native spirituality is the reciprocal relationship between humans and the natural world including the plant and animal nations and the entire cosmos. It cannot be overstated that Native spirituality is not merely a belief system; it is a way of life that incorporates this relationship as an essential and integral part of being Indigenous and doesn't preclude the use of empirical methods.

Because science is a human activity as well as a body of knowledge, it is influenced by a culture's epistemology. Western science developed within Western culture, which is rooted in ancient Greece, and a dominant characteristic is the bifurcation of life into sacred and secular realms. Western science is also limited to the material causes of events. These assumptions are deeply embedded in American life and are alien to Native worldviews. But it is important to acknowledge, as mentioned earlier, that this (materialistic) aspect of Western science did not always exist among the earliest Greeks, who embraced a unified view if reality—the unity of matter and spirit—instead of the dualism in current Western thought.

By the 15th century A.D., when European immigrants brought Western culture to this Continent, they had already abandoned the *original* Greek worldview which did not conceive of science, philosophy, and religion as separate entities. It was in the 5th century BC that a split occurred and science and religion went separate ways. As a consequence of this rift, Greek philosophers became occupied with questions about the spiritual world and the human soul, paying little attention to the material world. Aristotle, on the other hand, took an interest in the material realm and developed a scheme that became the basis for the Western view of the Universe.[75] Today, many Natives and others refer to this system of knowledge as Western science.[58]

What happened long ago in the Greek world was tragically forgotten in the West; Western civilization proudly claims to be rooted in Greek culture but the West is lost and oblivious to its holistic and spiritual roots. It has no spirit and, nowadays, science and religion are still at odds with each other; they are fighting twins, children of the same mother. Scientists typically claim that the former is connected to knowledge and the latter to belief, which is perceived to fill a human need but has no basis in reality. Indian religion is often viewed the same way, that it is buried in superstition and a human invention based on fear. In truth ignorance lies on the Western side.

The existence of different epistemologies is evidence of Mother Earth's human diversity. Western scientists want a science that applies to all situations, not realizing that knowledge among the Indigenous is *deep, complex,* and predominantly *local,* differing from place to place. Western thought limits science to the creative endeavor of the human mind, selecting from human experience only what can be counted and measured. It is mistakenly portrayed as the whole of reality.

---

[58] Some themes in the book appear more than once in different contexts. The origin of Western science is one of them.

Power and Ignorance

Western culture has created a powerful science from its effective use of that very old and widely accepted premise that denies the unity of matter and spirit; in fact, it denies spirit.

It is not difficult to see how such an assumption—in essence that we live in a world without spirit—can only lead to the idea of a dead Universe.

In view of the damage that scientific technologies have done to the landscapes within tribal homelands, as well as lands and sacred places that are still being desecrated, the existence of negative feelings toward science among tribal leaders is not surprising. It is a hard fact that science and technology were never intended to benefit the people who originally occupied this land.

Meanwhile, the deep knowledge of the natural world possessed by American Indians and other Native peoples has for centuries remained mostly unknown and unacknowledged by the larger society. As far as tribal knowledge is concerned, it is safe to say that most educated citizens perceive tribal life as unsophisticated and not up to par with Western civilization, or that lacking the advantages of scientific developments, it must inevitably be underdeveloped, primitive, backward, and still full of superstitions.

Instead of a source of knowledge, Indians are the subjects of research and the objects of Christian missions but their spirituality seems good enough to be appropriated by many who seek a more meaningful spiritual experience than what churches can provide. But there actually is no way that Indian spirituality, which varies from tribe to tribe, is connected with specific landscapes, and is not based on creeds and doctrines as in the church experience, can be properly experienced out of context. While it is good to acknowledge the spiritual hunger, it also seems appropriate to suggest that the first place one should look for satisfaction and fulfillment is in the roots of one's own culture.

This hunger finds expression not only in the appropriation of ceremonial activities but also in other ways. When I participated in a Sun Dance one summer in Montana, I found that non-Indians are good at being Indians, at least outwardly. I was surprised to learn that many non-Indians, who had learned to perform certain tasks well, were in charge of several parts of the ceremony: bringing in the cottonwood branches for the arbor, building the sweat lodges, building and maintaining the fire, fetching stones.

Almost all of the foremen were white men and some were bossy and authoritative. I was not a Sun Dancer but a "supporter" working inside the protected area. The non-Indians defended their freedom to dance, saying "This is my family" or "I'm walking the Red Road," but I learned that the Sun Dance was the *only* traditional Indian event they participated in all year.

According to Native news sources, at some of the Sun Dances there have been Indian elders who could not dance because they had been crowded out by whites who left no room for them on the Sun Dance tree.

Misperceptions

Misperceptions of tribal peoples are sometimes reinforced indirectly by conservative scholars. An example that represents widespread attitudes and perceptions is Phillip Johnson's *The Wedge of Truth*. Johnson is a legal scholar and a well-known advocate of creationism, who unintentionally denigrates tribal people while criticizing conventional science for its naturalistic, mechanistic philosophy. He states in that book that evolutionists are practicing idolatry because they use theories as idols, or as crutches, allowing them to deny the existence of God. He supports this statement with the remark that "primitive tribes make idols of wood or clay."[76] He states that from a naturalistic viewpoint "ideas that are classified as 'religious' are inherently tribal because they are based on subjective beliefs rather than objective reality"[77] to say that those ideas are not universally applicable.

Referring to the Trinity as a sophisticated doctrine that is not so easy to understand, he asserts: "Of course we would not introduce Christianity to African tribesmen or neglected slum children by reading them an academic treatise on the Trinity any more than we would introduce science education to the same people by giving them a lecture on the Copenhagen interpretation of quantum theory."

Quantum theory actually corroborates Indigenous worldviews and, as for the conventional Copenhagen interpretation, physicists have proposed three alternative interpretations to that interpretation.

The perception of tribal peoples as being unsophisticated and having nothing to offer others is apparently a widespread attitude in the West, as anthropologist Wade Davis indicates: "We have this extraordinary conceit in the West that while we've been hard at work on the creation of technological wizardry and innovation, somehow the other cultures of the world have been intellectually idle. Nothing could be further from the truth."[78]

Johnson then contrasts the intellectual sophistication of scientific and theological concepts against the basic story of the incarnation, which is a simple truth with a universal appeal that includes cultures that "want something better than the idolatry they've been given." He views Indigenous cultures not only as idolatrous but also as lacking in knowledge of their environment.[79]

He couldn't be more wrong. Although the book is not about tribal people, they are referenced generically to illustrate how evolutionists practice idolatry and, presumably living in remote areas ("primitive tribesmen"), don't have the intellectual or spiritual sophistication to understand theological truths or to possess a sophisticated cosmology.

In addition to the above troubling perceptions and attitudes, stereotypical images and public attitudes are perpetuated through television, the media, and film. Native educators and authors have identified other concerns about Western science including its dogmatism, its incompleteness as a unifying system of knowledge, its reductionist, unspiritual, and unsocial character, and its implicit and explicit attacks on tribal traditions. Scholars in Canada began in 1975 to "demythologize" the scientific fundamentalism that exalts Western science, according to Glen Aikenhead, emeritus professor from the University of Saskatchewan's College of Education, who has written extensively about aboriginal science education.[80]

Christian conservatives and many academic scholars from the Western tradition seem oblivious to the spiritual dimension that is central to life in Indigenous communities. Also troubling are the persistent misperceptions of Native peoples as primitive and backward. But countering them with "proofs" of Indigenous traditional knowledge is hardly appropriate or dignified; the truths of traditional knowledge stand by themselves. Native scholars would agree with the materialistic and mechanistic characterization of science but they view it as a failure within the Western tradition; the conflict between science and religion, or between evolution and Intelligent Design, does not occur in Indian country.

Land and Spirituality

This section is principally a personal message to the American conscience about the need for restoration in Indian country. As with the entire book, I am solely responsible for the thoughts, assertions, and logical arguments here presented. Please explore some ideas with me.

There is no doubt whatsoever about the existence of non-Indian individuals who care about Native issues, build relationships, take risks, and are generous with their time, efforts, and resources. It is the large-scale efforts which are lacking. However, the tribes are not waiting for the U.S. to take the initiative.

It is well known at least among Indians that the values and practices of the Western World have benefited industrialized nation-states while significantly damaging the ecological and social systems of lands long occupied and nurtured by Indigenous communities. Science and technology, science education, and research have ultimately had their effect on tribal homelands. But the tribes in the United States have not forgotten the solemn promises made to them when they granted rights to non-Indians in return for services.

The symptoms that plague Native America—alcoholism, poverty, dependence, violence, illness, suicides, breakdown of values, etc.—are the *effects* of abnormalities, but what are the causes? The same causal relationship applies to disease and treatment, that if the disease is cured, the symptom(s) should disappear. Some symptoms have multiple causes, but the principle of cause and effect still applies. Treating the symptoms brings some relief but will not cure the malady.

We need to know the factors that affect Native people and the principles that govern how those factors are related to the land and to the people's spirituality. Let's continue with some basic questions:

1. What is the source of America's wealth?

2. What is the source of hope for the nation?

3. What is the source of hope for Native America?

4. What represents the greatest loss to Native peoples?

5. What kinds of activities are best supported with money?

Consider also the following assertions:

6. Money is a human invention but not inherently evil.

7. The political democracy under which we are now governed and demands allegiance has no relationship to the earth on which we walk.

8. Land ownership is a human invention.

9. Real estate business and land development projects represent a deviation from the sacred because they use land as a commodity for profit.

These questions and statements may trigger reactions in your mind. All of them, including the first question, involve land. When European immigrants first arrived on this Continent, they had nothing, not even rights, only what was in their travel bags, and possibly some of the gold that had been stolen earlier from Meso-American tribes and had been transported first to Europe.

There were two waves of European immigration; those from Spain came first, then from England. English immigrants practiced trading for a while and eventually created money and banking. The Spanish not only stole the land; they also stole the people and used them as objects to produce wealth for themselves, not for the people.

We have to conclude that America's source of wealth is the land, all of which once belonged to the tribes but not in terms of private ownership. The tribes did not need money; they lived off the land and were primarily concerned with practical living and survival, not with materialistic wealth. American values, on the other hand, which are rooted in Western thought and ideology, are exceedingly materialistic, a fact that is reflected in the English language, as made obvious by the large percentage of English nouns that represent objects and things that satisfy, including those that are not necessary for a life of happiness.

Is there a principle connecting land to restoration and hope? It is Earth that supports life. Land and the elements— water, earth, air, and solar energy—form the Indian's natural/spiritual world. Spirit has always been here and everywhere. When the land is healed, the people will be healed.

But there is a gaping hole in the American consciousness about Indigenous knowledge and wisdom. Western values allow Mother Earth to be exploited for human gain; except for individual responses, which are relatively sparse, there is no respect for nature's authority despite the evident principle that what we do to the land we do to ourselves. America was built on an ideology that categorically denies Indianness: everything that is essential to the Indian: land, language, spirit, culture, family, community, and wisdom.

Indigenous traditional knowledge (TK) is not a noun, unless we think of it in English as a dynamic noun; we could say that it is a result as well as a process that requires living in, and relating to, a place and all that exists in that place for countless generations. It cannot be separated from the source. It is ethical, moral, spiritual, and sound from the standpoint of the West's best science—all because of Indigenous spirituality.

If the Creator placed the people on the land and gave them instructions; if they acknowledge and live in deference to the

Creator and his Creation, recognizing that nature has her own authority system of laws and principles; if they live in a reciprocal and respectful relationship to the land and to all things on the land so that nothing is taken disrespectfully and only what is needed is taken; if they live in harmony with the rhythm of the land and the cosmos; if ceremony and prayers of thanksgiving and other occasions by individuals and by the people are part of the flow of life in the community; and if parents and other relatives teach the children, through stories and practice and examples of generosity and wisdom, the skills they need as well as all of these other things mentioned, so that they learn to use their hearts and minds together from an early age; if all of these are true, then this is what, in my opinion, might be called "Indigenous spirituality." The land, the language, and the culture are inseparably linked together as the people travel along a self-sustainable, continuous, and endless course.

This is the process that defined the ancient world which our ancestors knew and cared for until it was abruptly interrupted by a people who were already alienated from nature in their own country where they had trampled upon the northern tribes of Europe. Restoring that sustainable world to a likeness of what was once experienced is not as simple as taking piecemeal legislative actions or adopting five- or ten-year plans that reduce fossil fuel emissions or replace them with renewable energies. *Only a change in worldview and lifestyle will have a noticeable effect.*

As to the original instructions, Kenny Ausubel, founder of Bioneers, explains that the journey to recover the sacred is what humanity needs most now and that the instructions partly reside in Indigenous Traditional Ecological Knowledge.[81] I don't doubt as well that some of the instructions have been guarded in the hearts and minds of elders from various tribes. In the handbook, *The Sacred Tree—* the product of a project that involved more than thirty Native elders, spiritual leaders, and professionals—we are told that the Creator planted a Sacred Tree for all the people of Earth to gather under and find healing, power, wisdom and security. It was foretold that the people would wander far away, forget about the tree's nourishment and protection, bringing great sorrow. We are told that the Tree itself would never die and that, through the ages, revealed truths have been preserved within the minds and hearts of wise elders and leaders.[82]

Historically, the Indigenous form of communicating and preserving many truths among the people has been through oral tradition. According to the Center for World Indigenous Studies, between 4000 and 6000 Indigenous nations still exist today. In the United States and Canada, they have suffered loss of language, land, and culture, but they thrive under the Sacred Tree, which contains all of the teachings that Creator has given to the Two-Leggeds, the Five-Fingered Ones (humans).

The impracticality of legislation to produce the desired results is obvious; and it is not the only problem. Blind conservatives in the U.S. Congress, influenced by greedy corporations, are not likely to cooperate, not to mention that to meet global crises head-on would require mobilizing all nation-states for cooperation.

Honoring the sovereignty of Indigenous nations also has a solid biblical basis. After Jesus left Earth, his apostle, Paul (Saul) of Tarsus, visited Athens, Greece where by invitation he met at the Areopagus[59] with Greek philosophers who asked him to explain his "new teaching" that contained "strange ideas." Paul stood up and said, "The God who made the world and everything in it is the Lord of heaven and earth and does not live in temples made by hands. ... From one man he made every

---

[59] A high court of appeal for criminal and civil cases

nation of men, that they should inhabit the whole earth; and he determined the times set for them and the exact places where they should live. God did this so that men would seek him and perhaps reach out for him and find him, though he is not far from each one of us."[83] Let's examine this passage:

The Creator (God) gave the world's tribes (*every nation*) their specific territories (*exact places*) for a specific time period (*he determined the times set for them*) because he wanted them to relate to him (*so that men would seek him*). Several tribal elders have said that their tribe was given sacred *original instructions* by the Creator, some of which are still told in stories. When the Creator, the only Sovereign, populated Earth with his (tribal) nations, he gave them rights and sovereignty to accomplish a purpose. By implication and according to American Indian tribes' own testimony, these tribal rights are *original* and *inherent;* they were not granted by any nation or other entity on Earth and have never been willfully surrendered. The future therefore rests on the fulfillment of Creator's original intent for his nations. *The world needs Indigenous knowledge.*

The desecration of these sacred endowments and instructions given to the nations by the Creator and the disregard for and violation of solemn treaty promises by the American people and U.S. representatives, made in the Creator's presence and in some case even in the presence of ministers of the Gospel, are therefore of great significance. The following paragraph summarizes the events as I understand them:

European immigrants took up residence on tribal lands that had been providentially granted to the tribes. They formed colonies, imposed war on the tribes, and eventually gained sufficient collective military strength to proclaim themselves independent from England, their mother country, after winning a war of independence. After establishing a federal government, they drafted a Constitution with the aid and influence of the Iroquois Six Nations (the Haudenosaunee) whose concepts of democracy were carefully considered. Driven by land fever and thoughts of glory, the new American nation continued to move westward, entering into hundreds of treaty agreements with numerous American Indian tribes, making promises under oath in the Creator's presence and in the presence of Christian ministers—promises that were dishonored —and eventually overcoming tribal resistance that meant an Indian depopulation rate of 95 percent or more, due to war and disease. After all this, no priority is given to restoring the honor and dignity of American Indian tribal nations.

While the Constitution defines an official relationship with American Indian tribes and tribal members, it also places the tribes in a subordinate position to the federal government. As a nation-state formed by conquest, the United States assumed its authority over many nations living within its jurisdictional boundaries. Originally, the United States had no sovereign status until it was recognized by the Haudenosaunee.

A fundamental difference between the Indian and Euro-American is that the Indian perceives and experiences a spiritual universe, one that is alive, with interconnected social networks, while the Euro-American conceives of a dead universe. The Euro-American sees external beauty in nature and is inspired by its beauty and majesty, whereas for the Indian it is a social and loving relationship.

The tribes cannot depend on America as their ultimate source of hope. The converse is true: America's hope, like the hope of all nations, depends on the healing of the land. This spiritual principle that recognizes humans as one strand in the network of interdependence that directly connects the hope of all nation-states to the knowledge, example and wisdom of

Native peoples, who, by the way, have never built weapons of mass destruction capable of annihilating other nations.

Considering that Christian America is expected to reflect a spiritual perspective on matters of war and peace, and in view of the influence of conservative churches on national policy, it is appalling that they deny a principle which applies to all nations and is clearly expressed in the Bible; the principle, as clearly stated in the Scriptures, is the following: A nation's blessings and survival do not depend on military strength. The psalmist, speaking in general terms, says: "A king is not saved by his great army; a warrior is not delivered by his great strength. The war horse is a vain hope for victory, and by its great might it cannot save. Behold, the eye of the Lord is on those who fear him, on those who hope in his steadfast love, that he may deliver their soul from death, and keep them alive in famine."[84]

Likewise, according to this principle, the hope of America or any nation will not ultimately be decided on the basis of military power. But unfortunately, it is safe to say, biblical teachings are commonly applied more in accordance with established lifestyles and ideals than by conviction. In any case, all peoples depend on the earth, and if the earth suffers, the people suffer. The laws that govern our bodies, the lives of creatures, and the life of Mother Earth were given by the Creator. To Native peoples, these laws are also spiritual and govern the proper relationship to the land on the basis of stewardship, not exploitation—a land ethic, as already stated.

A life of sufficiency

We must fundamentally change how we think about what we require for our happiness. Our society is conditioned to feel like we don't have enough—not enough rest, money, influence, feedback to our ideas, interaction, wealth, space, vitamins, and so on. We buy things that are unnecessary, always wanting more. We are bombarded by clever advertising that reinforces the idea that we *need* more. But many of us can learn to make better choices: either accumulate wealth for ourselves out of the things we don't need or re-direct them to higher causes. Each of these options has consequences but if we recognize the spiritual principle, we will know how to make the right choice. It is a simple idea but it has global significance.

It was Lynne Twist, author of *The Soul of Money*, from whom I learned the above principle, which she expresses as follows: "If you let go of trying to get more of what you don't really need, it frees up all that energy to make a difference with what you already have." She further states that this difference will expand.

Jesus taught this, too, in the parable of the servant who gained five more talents from the original five. If we transform our lives from a condition of always wanting more (scarcity, the delusion that we never have enough) to one in which everything we have is exactly what we need (sufficiency), we will have more resources and energy to work with the things we already have. It will lead to freedom from want and the ability to help others funnel their money in the direction of the highest causes for humanity. Money will always flow in the direction of a person's commitment, and an awakened consciousness can change that direction.

But that's not all. The money and energy that have been freed up and directed to a higher cause is not given as an act of charity, in my opinion. The anticipation of giving does something to our spirit. The spiritual experience begins at the moment of intent and anticipation even before the money leaves our hands; this is a *spiritual act*.

What was lost

I do not want to offend anyone by writing carelessly about the profound sense of loss that becoming separated from the land represents, a loss that is also difficult to comprehend. Stories are important because they expose the things that are sacred. Harold Napoleon, a Yup'ik man, tells the story of his people in Alaska. Deeply troubled by the problem of alcoholism and alcohol abuse among Alaska's Native people and the death of his own son, he wrote handwritten letters for four years to address the problem. In one such letter, now published as a book, he discounts the theory that Native people are biologically susceptible to alcohol, concluding that the primary cause is not physical but spiritual; the cure must, therefore, be of the spirit. He speaks of the spirit world in which the Yup'ik Eskimo lived, the world they knew before the influenza epidemic of 1900 struck the people, due to exposure to white immigrants.

For fifty years the Yup'ik people saw their family members suffer and die until eighty percent of them had perished. In their stories, they refer to this experience as the Great Death. The extraordinary force of the story can be felt only by reading the book. Mr. Napoleon describes how the people viewed their world as complete. "It was a very old world. They called it *Yuuyaraq*, the way of the human being ... Although unwritten, this way can be compared to Mosaic law because it governed all aspects of a human being's life."[85] He describes not only the physical and spiritual universe in which the Yup'ik lived but also *what they lost* when they had no choice but to accept another way of life.

Similar experiences occurred all across America as tribes were displaced during the settling of the West and during a thirty-year period known as the Indian wars. Broken treaties, land transfers, forced removals, contempt for ceremonial practices and spirituality, war, and disease belong to that part of history, creating a multi-generational condition that Native psychologists refer to today as the *soul wound* in Native America, evident by the high incidence of Post Traumatic Stress Disorder.

The spiritual world that Harold Napoleon describes points to the spiritual dimension that is lacking in the modern world of technology, exploitation, and an economy based on profit that not only accumulates wealth for a few, *out of the surplus*, and leaves others in poverty but is also used to support projects and policies that are destructive to the land. No bomb shelter or military power can save a people or a government indefinitely from its own vices.

Chapter
1 0

INDIGENOUS KNOWLEDGE

I f you are impressed by modern science and technology, you will likely be more impressed by what the power of the human being at one with nature can do.

The genius of the ancient Polynesians is illustrated in a story that Wade Davis tells in *The Wayfinders: Why Ancient Wisdom Matters in the Modern World*, which I can only summarize.[86] He describes a long voyage on the deep open sea that began by night on a vessel, a canoe called the *Hokule'a*, relying only on the master navigator's skills, applied to the full extent of his ability and discipline, his protégé's memory, and the fundamental elements of the Polynesian world: wind, waves, clouds, stars, sun, moon, birds, fish, and the water itself. The training to acquire such skills begins at the infant age of one, when he or she directly experiences the sea by being placed in tidal pools for hours at a time in order to feel and absorb the sea's rhythms.

The ancient Polynesians were not navigators, says Davis, but *wayfinders* whose objective is to find a chain of islands that will guide them to the destination. Clouds provide clues by their shape, color, character, and place in the sky. Each minute detail, as reflected in the nomenclature, not only allows a description, but more importantly, the crucial information needed for guidance. The tone of the sky indicates the humidity; a halo around the moon indicates light shining through ice crystals, and the number of stars within the halo reveals the pending storm's intensity; animals encountered are seamarks (as opposed to landmarks) that indicate the approximate distance to land; literally hundreds of stars and constellations, known by name, help to orient the canoe in the right direction; and all these signs (and others I haven't mentioned) have to be combined for calculations that are made continuously during the voyage. It's as if the vessel never moves but waits for the islands to arrive.

The voyage was conducted without modern instrumentation or the application of the equations of modern science, demonstrating that ancient knowledge was valid and effective long before the advent of modern science. Physical principles were known centuries before they were mathematically quantified into modern scientific disciplines. But this is not all; that knowledge includes *spiritual and possibly other factors which are lacking in modern science*. This is one of the clearest examples demonstrating the differences between Indigenous traditional knowledge and the science of Western civilization—a fact that also shows how crucial that knowledge is for survival.

The Polynesians learned skills and gained knowledge through experience by interacting directly with nature. In our modern day, operators depend more on instruments and less on interactions with nature; instruments detect and measure what nature provides; they mediate between nature and the human being —a function once performed by humans.. Academic courses teach principles that were learned and quantified over a lengthy historical process and were eventually incorporated into those instruments. But learning the principles of navigation through classroom work does not make one a navigator.

All of the data acquired in sequence during the journey on the *Hokule'a* are stored in the navigator's memory, telling not only where they are but also where they've been; if he loses the position relative to the reference course, the vessel is lost at sea, which is why the navigators go for 22 hours, undisturbed and without sleep.

Science and Indigenous Knowledge

The world's oldest living human cultures, rooted in Mother Earth, built and sustained their communities for thousands of years without relying on Western civilization for tools and technologies. Living with the land and patterning their lives in accordance with nature's cycles and in harmony with her laws, they invented adequate tools and designed structures suitably engineered for their local environments. Indigenous peoples' knowledge has always been independent of, and predates, the science of the Western World.

Indigenous peoples are often unintentionally portrayed by environmentalists and conservationists as helpless victims but not as sources of knowledge. They point out the threats to their livelihood and survival, such as the threat of being relocated and dispossessed. In a review of Mark Dowie's book, *Conservation Refugees*, the article's author expresses concerns about the rapid dwindling of the world's biodiversity, then laments the dark side of conservation efforts; namely, that Indigenous people are often forced out of their homes and into refugee camps, many times with tragic results.[87]

While it is true that Indigenous peoples are victims and the threats must be addressed, an important message that needs to be included, after saying that "tribes are no more invasive than lions" and that "their fate should at least be taken into consideration," is that Indigenous peoples *are themselves conservationists* and self-protectors of the ecosystems that are threatened. Readers need to know about the value of Indigenous knowledge as well as threats to the people and to the environment.

Modern scholars and the public at large may question the validity of Indigenous science, but as in the case of American Indian history, Indigenous knowledge has gone largely unnoticed and unexamined by the Western World, contributing to the misperception of the Indian.

It is reasonable to assume, as Native authors have also stated, that the word "science" does not exist in Indigenous languages. Native peoples live with Earth as their mother; they don't try to destroy it. It was not until they were invaded and colonized by European powers that their way of life began to deteriorate. The science and technology that originated in Europe and expanded to the United States was obviously not designed to benefit the tribes, whose lands were in large measure taken by force or by deceit to establish a new world order that does not respect the land.

This book advocates a science within a worldview that honors a respectful relationship with nature, the way it is supposed to be—and indeed it once was, until something happened that changed the course of history.

As mentioned in chapter 5, Native peoples across the continent have made significant worldly contributions that have been appropriated by conventional science,[60] including agriculture, architecture, astronomy, medicine, ecology, engineering, aquaculture, horticulture, and more.[88] With their conceptual and practical knowledge of the natural world they have built sustainable societies and structures that predate contact with Europeans. But because of the Western World's disinterest in tribal knowledge and tribal issues, it has not

---

[60] See the "Pre-contact science" section in chapter 5.

benefited from the proven science of the world's oldest living human cultures. Indigenous people from various regions of Earth are also familiar with weather changes through direct observations that they sustain throughout the year.

For example, the Haudenosaunee (Six Nations Confederacy, also known as the Iroquois), in their major address to the United Nations in 1977, reported climate changes among other issues.[89] The rest of the world did not heed these warnings. The Indigenous Kogi people on Colombia's Sierra de Santa Marta in South America (who call the people of the Western World their "younger brother"),[90] the Inuit in Canada's High Arctic,[91] and Natives in Alaska, have all been reporting warmer temperatures and changes in animal behavior due to climate change.

In view of the centuries of neglect by public institutions and the pressure for Indigenous peoples to adapt to mainstream cultures, Western scientists still do not recognize the validity of traditional knowledge. But adapting this knowledge into a Western context is not trivial. Whether attempting to convey tribal knowledge and philosophy pertaining to the natural world to Western scientific minds or describing relevant Western scientific concepts in terms that are familiar to Native audiences, either case requires an adequate knowledge of *both* worlds and the ability to communicate concepts authentically.

Only a small percentage of Indian men, women, and children survived disease and war as the United States expanded westward, causing a depopulation rate of nearly 95 percent as it acquired Indian lands. Instead of being recognized as a source of knowledge, they constitute the poorest segment of society, with an average unemployment rate of around 50 percent,[92] and they now represent a small percentage of the total population. American public education has failed them, having served mostly as a tool of assimilation instead of empowerment.

The knowledge and beliefs amassed by Indigenous peoples are the result of living and observing the Universe for several millennia—the same Universe that Western scientists have explored by meticulously recording and measuring information. This fact is largely unacknowledged by the larger society. In my study of quantum theory and relativity theory during the past decade, I discovered several consistencies between these physical theories of modern physics and long-held beliefs that form the worldviews of many American Indian tribes and other Indigenous peoples.

For example, the notion that the Universe as a highly interconnected system is a physics principle supported by quantum theory and general relativity theory; it is a conclusion inferred by the facts and principles (described in chapters 6 and 7) that comprise these theories, not one explicitly expressed in a single equation or equations. Physics has uncovered a principle of nature that agrees with the Indigenous precept that "all things are connected." Whereas one system of knowledge expresses concepts in sophisticated mathematical language, the other does it metaphysically using deep spiritual insight and personal participation with nature, albeit for vastly different purposes.

I have no doubt that in principle the world's scientists can embrace the sacred web of life without deviating from good science. An awareness of the sacred does not preclude testable theories. In fact, it should improve the integrity of scientific endeavors while preserving and increasing the human survival potential. But in view of the harm that scientific technologies have done to the landscapes on tribal homelands, the negative feelings toward science among tribal community leaders are not surprising. It is a hard fact that the European concepts and practices imported to the American continent were never intended to benefit the people who originally occupied this land. A new generation of scientists is needed who will restore and implement old wisdom, as expressed in Native traditional concepts and corroborated by modern physics, will challenge the Western paradigm, will embrace the sacred web of life, and will rely on spiritual guidance. We cannot wait for the United States or the Western World as a whole to change on its own.

Without outsider intrusion, Native peoples were able to sustain themselves and their environments through the ages; that they had their own science is without question, although the term "science" may not exist in any tribal language. Knowledge of their environments is cumulative and local. As to how far back in history this knowledge can be tracked, each tribe has its own stories and methods of retaining it. As for the Haudenosaunee, they tell us:

> People who are familiar with Haudenosaunee beliefs will recognize that modern scientific evidence shows that the Native customs of today are not markedly different from those practiced by ancient peoples at least seventy thousand years ago. Indeed, if the Iroquois traditionalist were to seek a career in the study of Pleistocene Man, he may find that he already knows more about the ancient belief systems than do the modern scholars.[93]

Indigenous knowledge is not the product of scientific "revolutions" representing rapid progress within a short time period. It is acquired in the normal flow of living for a long time in a single location in close relationship to the environment. The table below, derived from presentations by Native authors and speakers, lists some of the dominant values of Western and Indigenous cultures in relationship to Earth, or nature.

Western:	Indigenous:
Earth belongs to humans	Humans belong to Earth
Full dominion over nature	Preeminence of natural law
Modeled on linear thinking patterns	Modeled on cyclical behavior of
Tame the wilderness; civilize the	"Wild" is natural
Language of inanimate nouns	Language of spirit
Accumulate for profit	Give; take only what is needed

A careful examination of the values listed in the right-hand column, while pondering the fact that they have formed the basis of life for tribes from different regions of the world living close to their environments over the span of thousands of years, makes reasonable the conclusion that those values represent proven principles of survival and sustainability.

Most Native authors and educators in science fields refer to this science as Indigenous *traditional knowledge* (TK). TK and spirituality are woven into a single fabric of knowledge and philosophy. Thus, although TK, or Indigenous science, does not have the same connotation as Western science, it performs all of the vital functions needed to sustain human and non-human communities and their environments indefinitely. Their knowledge is unified in contrast to the specialized scientific disciplines of the West. It consists not only of *what* is known but also *how* it is known. Within the oral tradition, this knowledge is acquired *experientially* and transmitted orally, in contrast to information obtained indirectly from other, usually written, sources. This is a very important distinction, because it also implies that science is transmitted through the culture and begins at childhood. This means that learning is a life-long activity involving the entire community and the practice of "science" is influenced by the community's moral, ethical, and spiritual values. By the time a dedicated individual becomes an elder, he or she will have acquired much wisdom.

It should now be clear why a side-by-side comparison of TK and Western science is unrealistic and unwarranted. The Alaska Native Science Commission online article, "What is Traditional Knowledge?" further explains:

> The temptation to compare scientific and traditional knowledge comes from collecting traditional knowledge without the contextual elements. For example, Native people have a far richer and subtler understanding of the characteristics of ice and snow than do non-Indigenous people. In fact, some Native classification is available only by virtue of its relationship to human activities and feelings. These comparisons sometimes incorrectly lead science practitioners to trivialize traditional understanding.[94]

Western science offers many technological conveniences but ingenuity needs to be judged on the basis of wisdom, ethics, and morality. Human beings have the power to choose what to do with the things they receive.

Cataloging Indigenous knowledge

The categorization of knowledge into specialized areas, or compartments, is a Western trait. The Western mind likes to analyze, reduce, and fill gaps left by previous research efforts. This is not a value judgment, only an observation made in an attempt to explain the two separate and independent paths to knowledge that Native peoples and the Western World have followed.

Generally speaking, attempts to categorize Indigenous aspects of life would be difficult because life and learning are based on relationships, stories, and direct experience rather than academic activity.

Native people generally view the physical/spiritual world as a unified whole; for example, Earth does not reveal her secrets in specialized categories such as "geology," "chemistry," or "physics." From a tribal perspective, knowledge might be better categorized in terms of human skills (for example, demonstrating a high degree of intellectual ability or spiritual insight) but not in terms of academic disciplines. The need or urge to know does not stem primarily from intellectual curiosity but from the need to maintain subsistence lifestyles that reflect a concern for practical living and often the need to survive under extreme conditions. Unless their science worked under those conditions, the people could not survive. Viewing themselves as stewards, their traditional lifestyles also determine the kind of technology they need.

If these characteristics seem unrealistic, one need only to consider the vast difference between living in complete dependence upon the natural environment and living in a city where virtually everything is purchased locally or ordered: food, fuel, water, electrical power, housing, waste disposal, etc. When the people are forced to live this way and abandon their traditional way of life, the loss is significant. The spiritual loss in particular is unimaginable; spirituality is integrated into every activity. They do not attend church in order to get religion; their "church" is the land. Tribal community life requires complex social organizing and government. Tribal members have many responsibilities.

A traditional Indigenous response to the world, often mistakenly referred to as "religion," involves *the whole being* through experience, intuition, philosophy, active participation and interaction, and perception. The active relationship with nature is illustrated by the story that Kawagley and Barnhardt tell about a man from the arctic who is stranded on an iceberg. He has been cut off from shore and new ice has formed overnight between him and the mainland. Remembering the advice of his elders, he tests the ice by dropping his ice pick from a known distance. If it passes the test (the weight of the ice pick allows the point to penetrate but stops where it is attached to the wooden handle), the ice will hold him and he gets on the ice. "Although the ice crackled and waved, he drew energy from nature, and was in rhythm with the sea and ice, and coupled with lightness and buoyancy, made it safely to the other side."[95]

Let us imagine the man in the above example employing an academic approach to the problem instead of his traditional one. Using physics, he would build a mathematical model that would require, in this case, measuring the buoyant force of sea water, which depends on the density and volume of the water displaced, comparing it with his own weight and the ice, and estimating the speed and length of the wave. Then, in principle, he could have synchronized his movement with the traveling wave as he walked on the ice. Obviously, this method has too many unknowns and uncertainties, making it too risky and time-consuming to be depended upon for saving his life.

Indigenous Metaphysics

American Indian reality includes *philosophy* and *experience*, forming a coherent view of the world. An Indian relates to the cosmos personally and socially. The late Vine Deloria, Jr. expressed this relationship as follows: "The best description of Indian metaphysics was the realization that the world, and all its possible experiences, constituted a social reality, a fabric of life in which everything had the possibility of intimate knowing because, ultimately, everything was related."[96] While metaphysics transcends the physical realm in search of ways to explain phenomena, tribal beliefs nonetheless stem mainly from empirical observation. It must also be understood that, for the Indian, no experience is discounted just because it can't be replicated; this may represent knowledge but not necessarily science in the Western sense.

Metaphysics is a branch of philosophy, dealing with the nature of reality, which seeks to explain phenomena that are not subject to observation, analysis, experience, or experiment. But Indigenous science, as I understand it, is not simply a combination of Western science and metaphysics. It involves Indian philosophy, which non-Indians often mistakenly perceive as Indian religion or superstition. Tribal metaphysics acknowledges a Universe that is alive; it transcends conventional science by not relying exclusively on measuring techniques for obtaining and validating information, and it does not display the sterile attitude toward the natural world that is typical of the West. As the story of the man on the iceberg demonstrates, the Indian's response to the world *involves the whole being* through experience, intuition, philosophy, participation and interaction, perception, and so on. Concepts that arise in modern physics, such as networks, relationships, flux, energy, wholeness, and nonlocal connectedness, are also part of the Indian's world.

From the time of Plato in the 5th century BC and through the Middle Ages, the study of nature was called "natural philosophy" or "Greek science." During the rise of modern science there were attempts to rid science of metaphysics, in part because natural philosophers were drawing conclusions without proof.

Even if Western scientists recognized the validity of Indigenous knowledge and were open to considering Indigenous principles for possible inclusion in the Western paradigm, adapting it into a Western context is not a trivial matter. An alternative approach is for the teacher to assume the role of "culture broker" by bringing traditional culture-bearers in contact with learners to teach or explain Indigenous concepts.

This is a method that Glen Aikenhead, professor emeritus of the University of Saskatchewan, has used with students in Canada. Based on many years of experience, Aikenhead has co-authored a book that is specifically designed to address the need for bridging Indigenous and Eurocentric cultures and is of particular value to science teachers and administrators. [97]

The verb-rich nature of Native languages corresponds to cultures that emphasize action and movement. So it is not surprising that a culture whose language reflects movement also has a language that emphasizes action, such as dancing, singing, and other ceremonial activities. It seems obvious that such a language can also describe the invisible events occurring in the depths of nature.

Indigenous traditional knowledge includes sense experiences[61] from the spiritual realm and it does not separate the natural from the spiritual. Why not then expunge the dominant biases and attitudes that motivate much of the world of science and bring *legitimate* facts and theories into contact with Indigenous metaphysics in order to reveal a unified world that includes realities from the Indian world? Such an endeavor would help restore Native wisdom, dignity and honor while manifesting the very principles that science has needed yet scorned since the beginning.

My premise is that the knowledge of Native peoples is in agreement with tested laws, particularly those in modern physics. Native peoples have long observed and lived in deference to the same Universe that people in science have explored and from which they have deduced laws based on recorded and measured information obtained impersonally. But before we go any further into discussing differences, we must remind ourselves that American Indian tribes were given

---

[61] A term used by Einstein with reference to information provided through *direct* sense experiences, which are psychic experiences of a special kind.

The Condor and the Eagle

original instructions to care for their lands and people. The Indian experience includes and extends beyond the natural sciences but it also emphasizes skill and intuition in addition to empiricism and it is driven by a sense of purpose and a different vision compared to the West.

I believe that the places where the two systems of knowledge meet can become gateways to realms that are unfamiliar to the Western world. It is Western science that needs to become re-aligned for the sake of all our relatives, and it is the West that needs to begin embracing *all* of our relatives.

Perhaps Western scientists are willing to listen. In February 2003 a panel of faculty members, including myself, from different colleges and universities in the United States presented a symposium at the annual meeting of the American Association for the Advancement of Science (AAAS) in Denver on the topic "Science Education in a Tribal College." We also had a special session with the press. This historic event, organized by Roberto Gonzalez-Plaza from Northwest Indian College, was described by one science writer as a "breakthrough" for Native science.[62] In 2008 we returned to AAAS and again gave presentations.

What seemed to be emerging in 2008 was not a new "movement," as some have suggested, but an increasing awareness of the valuable work by Native scholars who have been conveying to the Western world the validity of American Indian traditional knowledge (TK). In all disciplines, good science involves persistent effort, verification, and diligence. In this regard, TK spans hundreds, even thousands, of years of observation, validation, skill, and practice by the world's Native

---

[62] Science writer Richard Simonelli has contributed many articles to *Winds of Change* magazine, a publication of the American Indian Science and Engineering Society (AISES). "Native science" refers to the ancient knowledge of Native peoples. A book by Gregory Cajete, Ph.D., also bears that title.

peoples—and it represents a much older perspective than a few human generations including the entire history of formal education.

The lack of world-famous scientists among Native people is due in part to the scarcity of Indians, the scarcity of college degrees among Indians, and the scarcity of Indians interested in academic science. Besides comprising only about two percent of the U.S. population, there is a small number attending and earning degrees from colleges and universities. Indian contributions to the world are not channeled through networks and infrastructures that bring fame. But fortunately, many American Indian contributions that precede modern science and even pre-contact science are catalogued in Keoke and Porterfield's encyclopedia, containing more than 450 entries, which was cited earlier. [98]

Aside from the credit deserved by Native peoples, it is important to maintain a traditional perspective by giving significance to the ultimate consequences of human ingenuity— whether it has created a better world, not in terms of conveniences but in terms of peace, happiness, welfare, and sustainability.

Comparatively small numbers of Indians have earned graduate degrees—or any degree. A glance at history reveals that the college and university experience is relatively new in Native America. John L. Phillips quotes the following remarks by Secretary of Interior Caleb B. Smith in 1862 at the time of the First Morrill Act:

> "...participation in higher education by American Indians was almost nonexistent by the 1960s. In 1961, only 66 American Indians graduated from a four-year institution. A decade later, the post-secondary attrition rates for American Indian students reached 75%."[99]

The Condor and the Eagle

Indigenous Cosmologies

It is informative to compare aspects of physics and tribal beliefs that relate to cosmology. Here we will look at three aspects: Bohmian concepts that agree with or correspond to a Native paradigm, tribal beliefs that relate to physical cosmology, and Indigenous cosmologies.

<u>Bohmian Concepts</u>

With regard to the first aspect, physicist Bohm, a colleague of Einstein, expressed concerns for many years about the fragmentation that exists in science and society, prompting him to assert that a new, non-fragmentary worldview was needed. He did not embrace the conventional Copenhagen Interpretation of quantum theory and formulated his own cosmology based on the premise that the Universe is a unified whole. He referred to the natural world in terms like "objective wholeness" and "the undivided Universe," the latter of which is the title of his last book, which he completed just before his sudden death in 1992; it includes all of the supportive mathematics.

Bohm postulated his own unique model of reality, a cosmology that includes tenets which are closely aligned with those shared by many tribes. He was primarily motivated by the incompatibilities between quantum theory and relativity theory. He did not accept complementarity as the explanation of wave-particle duality or other aspects of the Copenhagen Interpretation. I have summarized his cosmology in the following beliefs and propositions:[100]

- A new, non-fragmentary worldview is needed in science.

- Physical phenomena are reducible to fundamental particles and laws describing the behavior of particles, an approach to quantum mechanics which therefore is ontological, not epistemological.

- The *implicate order*: The Universe is an unbroken, coherent whole. This wholeness is called the "implicate order" — a model of reality that can be compared to a cosmic web of interconnectedness or a hologram, a two-dimensional surface, such as a photographic plate, every region of which is encoded with a three-dimensional image that can be reproduced with the aid of a laser. A mathematical function called the *quantum potential* fills all space-time in the implicate order, is assumed to exist beneath space-time at the subquantum level, cannot be observed, and fulfills a requirement for non-locality. The trajectory of a particle is due to the underlying quantum potential. The undivided whole is primary, encompassing all, and the implicate order is inherent within the whole, not parts of the whole, like particles, quantum states, and continua. The whole encompasses all things, structures, abstractions, and processes. The explicate order represents the classical world of elements that we can sense at the macro level in our everyday life; the implicate order is what lies at the micro level, beneath space-time.

- The explicate enfolds into the implicate, which unfolds into the explicate in a continuous movement, a process called the *holomovement*.

- Modern languages (including English) are not suitable for describing quantum processes. (But as Bohm discovered, Algonquian and other Native languages *are* suitable, being rich in verbs.)

In addition to the above, there are several other concepts from physics, described in this book, which correspond or are compatible with a Native paradigm: The interconnectedness,

The Condor and the Eagle

interrelatedness, and interdependency of all things; the observer/observed relationship which implies subjectivity in the process of acquiring knowledge; everything is in motion, which is the source of all energy; the ubiquitous cycles in nature; Earth and Universe as a single living system; wholeness (e.g. emergent properties, non-locality, etc.) which is evident in systems and subsystems; and more.

Tribal Beliefs

The following beliefs are shared by many tribes:

- All things are interconnected, interdependent, and interrelated, belonging to a coherent whole; *i.e.,* "we are all related."

- All things are imbued with spirit.

- The source of matter, from an Indigenous perspective, is spirit; *i.e.,* spirit manifests as matter. (From a physics perspective, energy manifests as matter, since energy-to-particle and particle-to-energy transformations commonly occur in nature and in the laboratory.)

- The cosmos is always in a state of flux; it is constantly in motion and always changing, or becoming.

- Renewal and self-organizing processes occur spontaneously in nature; tribal ceremonies celebrate the renewal cycles of nature.

- The tribal relationship with nature is mutual and participatory, involving subjectivity.

- Everything belongs to a cosmic web of relationships and has the potential of intimately knowing.

- A Power, often referred to as the Great Mystery, permeates the cosmos.

- The power in nature works in circles (cycles)

- Matter is equivalent to vibrating energy (a Mayan belief).[101]

Indigenous Cosmologies

For the third aspect of comparison, I list the cosmologies and other characteristics of various tribal groups. The ones below are *specific* in each case but may be more widely shared.

1. Indigenous (Native) languages are action oriented. These languages are rich in verbs, implying action and movement. Some Indigenous languages either have no nouns or pronouns, or they facilitate speakers to express themselves without nouns. Bohm noted that modern languages, including English, inherently fragment reality because of their subject-verb-object structure, in that order.

2. Native languages reflect spirit and dynamism. They describe a world in motion, consistent with the actual behavior of the cosmos. According to Joseph Rael (Picuris Pueblo and Ute), in the Tiwa language, Tiwa words speak of a dynamic unfolding reality that is constantly ongoing. The Tiwa language has no nouns or pronouns, he says, "… so things don't exist as concrete, distinct objects. Everything is a motion and is seen in its relationship to the other motions." Further emphasizing the dynamic character of the language, he states the following:

> "For instance, a cup is *tii*, but *tii* means crystallized awareness or awareness that is in the process of crystallizing and uncrystallizing. That cup is not fixed. In Tiwa it is a continual unfolding of *tii*, and *tii* means 'the essence of the power of crystallization that is influencing awareness" and that awareness is the

The Condor and the Eagle

awareness of holding something in, like a spoon, or like tea or coffee. That coffee or tea that is in the cup is not static, either, but it's also in continuous motion." [102]

3. *The word "knowledge" is more than a noun.* Because there is no corresponding word in English that correctly translates the Native word for "knowledge," authors Aikenhead and Michell in their recently-published book, *Bridging Cultures: Indigenous and Scientific Ways of Knowing Nature,* introduce the expression "Indigenous ways of living in nature" (IWLN) as a replacement term for "Indigenous knowledge."[103] A key point is that the English term, "knowledge of nature," does not translate easily into most Indigenous languages, which are action-oriented.

4. Traditional knowledge is attached to the resource as part of the same biodiversity. It is not possible to dissociate traditional knowledge from the resource because the people and the natural resources they have cultivated both possess that knowledge,[104] a principle expressed by the Indigenous peoples of Colombia.

5. Language comes from the land. Jeannette Armstrong (Okanagan), in *Sharing one Skin: Okanagan Community,*[105] says that "The land taught us our language. To know all the plants, animals, seasons, and geography is to construct language for them. … [O]ur most essential responsibility is to bond our whole individual and communal selves to the land." She describes how the first, second, and third parts of an Okanagan word speak, respectively, of the physical realm, the unseen (mind, spirit, intellect), and the fact that the people are tied to everything else.

6. Walking into the tundra meant entering into a spiritual realm, says Harold Napoleon (Yupik), in *Yuuyaraq: The Way of the Human Being,*[106] In that world, the Yup'ik and the spirit beings were alone. It was a complete world.

Yuuyaraq, which means "the way of the human being," governed all aspects of human life. The visible world was only one aspect of *being*. The unseen world was the spirit world.

7. Other prehistoric cosmologies: the Chacoan culture of the American Southwest, the Nazca culture of Peru, the medicine wheel in Wyoming, Ollantaytambo and other megalithic structures in Peru, and more.

Chapter 11

THE EFFECTS OF COLONIZATION ON SCIENCE

We went to Geneva, the Six Nations [Iroquois], and the great Lakota nation, as representatives of the indigenous people of the Western Hemisphere; and what was the message that we gave? "There is a hue and a cry for human rights," they said, "for all people." And the indigenous people said, "what of the rights of the natural world? Where is the seat for the Buffalo or the Eagle?... We are indigenous people to this land. We are like a conscience; we are small, but we are not a minority; we are the landholders, we are the land keepers; we are not a minority, for our brothers are all the natural world, and we are by far the majority."[107]

During my entire educational and most of my professional experience, I did not question the assumptions of the West, which also form the basis of its science. But upon discovering my tribal identity, I consciously decided to re-examine my entire educational and spiritual experience under a different set of assumptions.

Western thought treats the spiritual realm as a separate and non-essential compartment of life. It perceives a universe that is dead and without spirit. It treats the earth as a dead commodity that exists for the sole purpose of exploitation by humans and perceives the natural world as having no rights. The science that is built upon it relies exclusively on the physical senses for information, which is considered the only valid basis of knowledge. In reality, it limits our understanding of the relationships that exist in a universe of matter and spirit and can provide no more than an incomplete and fragmented view of the world.

On the first day of high school biology, Mr. Lucas, our biology teacher, wrote a word on the blackboard at the beginning of class: "polymorphonuclearneutrophilicleucocyte." I didn't know what it meant but I think he either wanted to impress us or tell us that this compound word represents how different parts of biological knowledge are connected.

Indigenous worldviews perceive the universe as a unified whole. Unless conventional science sheds some of its priesthood character of narrow specialization and begins to think in terms of unified wholeness, several problems in science will remain elusive. Because American culture is rooted in Western culture, we can also expect to see a highly compartmentalized and fragmented American society, which I believe is evident.

How might have American society benefited from allowing Indians to incorporate their knowledge and practices into the American experience? Science itself suffered, albeit indirectly, from what it was not allowed to become: a system of sustainable principles and practices. For many tribes, the traditional practices that once supported a subsistence way of life became only a memory. Indian customs and language were either suppressed by federal Indian education policy or criminalized by the government. Science was not allowed to develop within the context of a holistic worldview. European immigrants failed to see the hidden wisdom of Indian traditional ways that care about the whole human being in relationship to the natural world. Instead of the spiritual health that once prevailed in

Indian communities, a standard of living based on materialism, politics, overindulgence, and greed characterizes modern society.

Native educators have been saying that Indigenous science is not just for Native people; it is needed to address the problems of this century. Our relationships to Creator, to ourselves, to the natural world, and to other human beings and nations are suffering. Restoring balance will require a radical change in thought and practice, including terminology that reflects the current paradigm with reference to the natural world; we speak of the "environment" as if humans were separate from the nature that sustains them.

Native elders from all regions of the Continent collectively tell us that humanity must awaken to the urgent need to care for Mother Earth and all life. The profound changes that have occurred to the people and the land should prompt a spirited demand for changes in science. One encouraging fact is that Native educators are bringing traditional knowledge in contact with Western science through science curriculum planning, publications, and research that reflect Indigenous worldviews.

Widespread mainstream attitudes toward nature are still primitive but some positive signs of change are evident. Western scientists have begun to use more holistic approaches to address environmental problems. We are also witnessing a shift from a science of specialization that reduces facts into seemingly unrelated fragments to a science that recognizes the more realistic complexity of self-organizing systems at work in the natural world. Both of these trends overlap with Indigenous worldviews. Although neither of these trends is motivated by an awareness of traditional knowledge, they nonetheless represent changes toward holistic science.

The need for a new paradigm in science is also recognized by non-Native educators, who are building appropriate curricula for Native students in collaboration with tribal community leaders and elders. Other researchers are spearheading practical research on the nature of reality that acknowledges the inherent wholeness of the body and mind.[108]

I believe that decolonizing efforts should also involve explaining some of the relevance and inadequacies of conventional (Western) science, showing how science and science education can and must change, and emphasizing the validity of Indigenous traditional knowledge. This approach may attract Native students, in particular, to the study of science (or physics, as I have described in this book) and demonstrate to community leaders and educators that science can change through exposure to Native ways of knowing.

As a Pueblo Indian, I did not grow up in the traditional way but I have associated and collaborated with Native faculty who are connected with their communities and know their traditions. Some of my learning was possible through the help of many teachers as well as my own self-taught ways that have engaged my whole being, including active participation in the ceremonial and social life of tribal communities, interactions with Native colleagues, listening to elders, teaching physics and history to American Indian students, directing an environmental studies program at a tribal college, prayer and meditation, arduous reading and study, and activism. But nothing can give us what can only be learned along the path of life's journey: our own language, the songs, and the stories that are best learned during youth.

When I was environmental studies program director for a tribal college, I traveled and attended tribal group meetings around the Northwest and Alaska to describe and promote our two-year degree program. Hoping to make my presentation interesting and relevant, I emphasized several facets of our program which was designed specifically for Indian students:

our motto, "come as an Indian, leave as an Indian," which emphasized how the sequence of classes was framed around Indigenous themes of land, air, and (inland and ocean) waters; the sense of community due to the mutual support that students gave each other, particularly during difficult times; the interactive classroom atmosphere; the outdoor learning workshops; and the dedicated faculty who not only taught courses but also designed the program.

It was also obvious during interactions with some tribal leaders that science did not play an important role. Some comments were very direct, indicating a strong disapproval of science education and possibly because I was promoting it: "We don't need science!" Those feelings toward science were not surprising, considering the harm that Western technology has done to the ecological and social systems of lands long occupied and nurtured by Native communities.

Western thought separates science from theology and spirituality, calls it "religion," and places it in a separate, more sophisticated, domain. Indigenous knowledge systems, on the other hand, do not practice or conceive of such a separation. In any case, helping students appreciate nature and enjoy the study of the natural world is the responsibility of all educators.

Several terms in common usage do not quite correspond to the conceptual world of Native peoples. "Study" is one of these, which for Native peoples is better called "learning"; it is integrated with life in the community rather than a separate activity.

Because the conventional classroom approach to learning that is associated with formal education is also out of context with the natural world, it has created the need to develop relevant curricula for Indian students that incorporate aboriginal values and aboriginal ways of learning. The classroom approach feeds the mind; the traditional way works through respectful relationships with our non-human relatives living in their natural environments; this is the Indian meaning of "community." Our relatives have a consciousness of their own that must be respected.

Culture has played a significant role in public education, including science. Much of the information we absorb daily is conveyed through the culture of the dominant society, whose values also influence us through the news media, television programming, church, and other ways. A society's language also reflects its culture and has become the dominant one in many homes where other languages, including Native, are spoken. Eventually, we develop our own foundational philosophy, or worldview, that operates subconsciously without prompting whenever we need to form a judgment or make a decision. Within this worldview, we contemplate the world, interpret experiences, and assert what is real without thinking about the assumptions that are deeply seated within us.

Science and Culture

One common misconception is that Western science represents truth, that it consists of verified and incontrovertible facts. This can't be true because science is not only a body of knowledge but also a human activity. Furthermore, knowledge relies on interpreted facts and is therefore influenced by culture. A culture's theory of knowledge, or epistemology, encompasses the doctrines and assumptions of that culture and addresses such questions as what it means to know and how knowledge is validated. Different assumptions will lead to a different science.

The deeply-engrained notion that knowledge exists only if it can be measured is associated with the mechanistic and reductionist philosophy which asserts that reality is purely material and that the Universe is reducible to its most basic physical constituents or "building blocks." A related notion is that the Universe can, in principle, be disassembled and re-assembled like a machine whose parts work the same by themselves as when they are connected to form the whole system. A cultural trait associated with this reductionist philosophy is the separation of life into sacred and secular domains. Today, as is well known, knowledge is highly categorized into separate compartments of specialization—another example of reducing and cataloging information into smaller parts.

These concepts are rooted in Western culture. Expressions of pride commonly appear in American textbooks about the accomplishments of "our" Greek ancestors, who of course are not the ancestors of non-Western peoples. Were it not for a historically significant rift that occurred among the ancient Greeks in the 5th century BC over the long-held principle of the unity of matter and spirit, Western science might have developed holistically. Prior to that time, according to Fritjof Capra and others, science, religion, and philosophy were not treated separately.[109] The Greek philosophers from that era initiated a path that has now been in existence for more than two thousand years. Aristotle developed a scheme that was first known as Greek science, survived through the Middle Ages, and became the basis for the Western view of the Universe.

By the fifteenth century, when Western culture was imported to North America by European immigrants and later developed into American culture, it had already deviated from the original holistic worldview of the Greeks. Science and religion remained separate and often conflicted with each other, with scientists typically claiming that the former is connected with objective knowledge and the latter with subjective belief. Some scientists view religion as irrational and superstitious, completely disjoint from science. But if Greek science had developed holistically, perhaps the clash between Western and tribal cultures would not have been as severe and the violence that ensued as European immigrants moved westward, scorning Indian ways, might have been averted.

Spirit/matter dualism persisted until the seventeenth century, when René Descartes devised a philosophy that further split nature into two separate and independent entities of mind and matter, where matter is completely dead and the mind is separate from the body. Further fragmentation caused humans to separate themselves from the natural world, commonly referred to today as the "environment." This mechanistic philosophy, or Cartesian dualism, took hold and became the basis for a highly successful and powerful science, possessing the ability to express many physical principles in mathematical language. But mathematics per se is not to blame, nor does it preclude holism.

The existence of different epistemologies and knowledge bases are evidence of the earth's human diversity. One point of divergence between Western and Indigenous thought is the scope of experience that is considered valid. Western thought limits science to the creative endeavor of the human mind, abandons the rest of human experience, and portrays this incomplete knowledge as the whole of reality.

Traditional Knowledge

In the traditional life of Indigenous peoples, knowledge transcends the physical domain and incorporates the whole of experience. It recognizes that humans are spiritual as well as physical beings. With or without awareness, since all things are connected, everyone is related to the rest of the cosmos—a principle that was not learned through dogma. When there are no presumptions and the Universe is viewed both as a laboratory and a place of mystery, additional paths of discovery become possible and every experience is given significance. The reciprocal relationship between humans and the natural world, which includes the plant and animal nations and the entire cosmos, is an essential element.

In a holistic worldview, what is commonly referred to as scientific knowledge, meaning knowledge obtained using the methods of Western science, is only a part of a unified whole. Science and religion are no longer conceived as separate entities and the Cartesian split of spirit/matter or mind/body is no longer assumed; however, this alone does not imply that we can immediately formulate a new science.

Care must be taken in how we think of science in the context of what Native peoples know and how they live, which is often referred to as traditional knowledge (TK) but involves more than science in the Western sense. One aspect of TK is the cumulative knowledge amassed by Native peoples through centuries or millennia of living in their environments and is not limited to the physical domain. Another aspect involves *how* this knowledge is obtained and how the people live as they learn. "Knowledge is a collection of facts and information stored in memory. It implies a focus only on **what** is known and separates it from **how** this is known or the true context in which something is known."[63]

Science makes heavy use of mathematics, symbols, abstractions, analyses, indirect methods of measurement, and so on. The empiricism in TK, on the other hand, is inherent in the traditional way of life, which takes into account all phenomena within a comprehensive view of the world.

A science that is limited to the physical domain needs a source of moral and ethical guidance. Its service to humanity

---

[63] This quote [emphasis is the author's, not mine] is from a document which Patricia Cochran, who at the time (ca. 2001-2002) was executive director of the Alaska Native Science Commission, distributed during a planning meeting in which I participated.

will depend on guidance from a higher form of knowledge. In a holistic approach, on the other hand, scientific endeavors can be guided by the influence of consciousness, conscience, and spirit cooperating in community. A reciprocal relationship recognizes nature as an authoritative system that implies human responsibility. Scientific knowledge in itself has no moral value.

The ability to make moral choices in science is an issue whose resolution may be best illustrated with a story that Vine Deloria re-tells in *Evolution, Creationism, and Other Modern Myths*.[110] In 1918, a Christian missionary by the name of A. McG. Beede took Harry Boise, a Yale graduate, to the Standing Rock Sioux and Turtle Mountain Chippewa reservations, where Boise explained the scientific method. Beede later wrote in his report that both groups immediately understood the concepts without difficulty. A Sioux elder by the name of Red Thomahawk reported to Beede the following consensus: "The knowledge and use of any or all the powers of the objects on Earth around us is as liable to lead a man wrong as to lead him right. It is merely power, with no way of knowing how to use it correctly ... unless Woniya [Spirit] is with a man's spirit for the light." Chippewa leader Rising Sun gave a similar response: "The 'scientific view' is inadequate to explain, among other things, how man is to find and know a road along which he wishes and chooses to make this said progress unless the great Manitoo by his spirit guides the mind of man, keeping human beings just and generous and hospitable."

As Deloria explains, these two tribal groups placed ethical and moral knowledge above any knowledge of the physical world, which is the *only* knowledge that purely empirical methods can provide. Moreover, this higher knowledge is accessible. Oglala Lakota medicine man Black Elk said, when he spoke about his great vision, that he used the power from the Outer World to cure the sick and that with the visions and ceremonies he was only a hole through which the power could come. We can choose to harness the power of nature for the wrong or the right reasons. The lesson from this story is not a physics principle but a *spiritual* one. Humans have the freedom to make choices, but to make the right ethical and moral decision, the intellectual powers of the mind alone are insufficient; we need light from Spirit.

The above story also exemplifies the role of subjectivity in relationship to knowledge in that acquiring a nobler kind of knowledge depended on guidance by Spirit. Knowledge is a subjective experience involving consciousness, which consists of experience and reflection. Reflecting on what we think about (*i.e.*, the content of our thoughts) gives us a perception of reality. Metaphysics is neither science nor theology but, as has been suggested, it can serve as the arena for common reflection.[111] Phenomena do exist that can neither be proved nor explained. Despite its power, the scientific method is also *self-limiting* in scope and not equipped to explain the totality of human experience.

It is ironic that Native peoples, who have a wealth of knowledge and wisdom and even an aptitude for science, have always faced problems of survival since contact with Europeans instead of being recognized and embraced for who they are. The United Nations (UN) officially recognizes their value to the world and their right of existence. Recognizing the threat of extinction of the earth's peoples, the UN in 1948 unanimously adopted the Convention on the Prevention and Punishment of the Crime of Genocide. According to the Center for World Indigenous Studies, there are some 3,000 Indigenous nations (peoples) in the world today, compared to less than 200 nation-states (states with nations living within their territories) within the United Nations. More than 550 federally recognized tribes (nations) live within the territorial boundaries of the United States, a nation state.

The article, "What is Traditional Knowledge?" by the Alaska Native Science Commission describes the nature and immensity of knowledge accumulated by Native peoples over centuries and millennia of living in close relationship with nature.[112] In parts of Canada, TK is often referred to as aboriginal knowledge; it requires direct observation of how each ecosystem functions, the proper management practices and techniques required to sustain them, and a deep understanding of appropriate relationships to the plants and animals. TK has to be reliable, because the people depend on it for survival. They do not simply explore something "because it's there"—an important difference between Western and Native science.

Because ecologies vary from place to place, different knowledge systems exist throughout the world. These systems are vulnerable to external pressures, such as intrusions by other humans or governments seeking to extract resources. The Cross-Cultural Science & Technology Units Project at the University of Saskatchewan in Canada and the Alaska Rural Systemic Initiative at the University of Alaska are two examples where systemic integration of Indigenous and Western scientific knowledge is occurring.

An important aspect of traditional knowledge systems is that the biodiversity of a particular place (*i.e.*, the range of organisms present in a given ecological community) is the collective result of transformations created over time by the people's own science across generations. Lorenzo Muelas Hurtado from Movimiento Autoridades Indígenas de Colombia explains that the Indigenous peoples of Colombia do not dissociate traditional knowledge from the resource.[113] It is not possible because the people and the natural resources they have cultivated both possess that knowledge; it is attached to the resource as part of the same biodiversity. Unlike the manner in which Western scientists acquire and catalog knowledge into compartments through specialization, Indigenous peoples generally view the world as a unified whole.

A conscious awareness of the unseen world of spirit and respect for the powers in the universe characterizes the spirituality of Indigenous societies. Spirituality is tied to the specific homeland, the familiar landscape that nurtures the tribe, where traditions, customs, teachings, beliefs, prayers, ceremonies, and language form a complete and harmonious world. It is not an abstract notion, political ideal, or set of doctrines that, for many in non-Indigenous societies often bring the comfort of belief at the expense of the continuous pursuit of knowledge. All living things are seen as relatives. The people view their role in a covenant and reciprocal relationship with the land as stewards and guardians of that part of the biosphere that they occupy. Traditional knowledge is the peoples' own science; it is a way of life. The knowledge systems are unique to the Indigenous societies living within their environments. The people know they belong to the land.

In industrial societies, including America, undue reliance on technologies tends to suppress the use of human powers. Unlike living in a world of affluence and modern conveniences, Native people have had to acquire skills in order to survive the elements, such as the inhabitants of the Arctic regions who have learned to predict the weather before going on dangerous hunts that last several days. Native science (traditional knowledge) gives the ability to arrive at solutions without complex mathematical models.

I believe that Native physicists can, as I do, identify spirit with the energy that pervades the universe without difficulty. As spiritual beings, all of us sense this energy in everyday life. One of the tenets of the Maya is that the spirit is manifested in the soul; they perceived long ago that the soul is the manifestation of spirit, which according to Hunbatz Men, is intellectual energy. Modern physics makes this notion even more credible, since transformations between matter and energy are always occurring. If the soul perceives the mind, which is not material, then energy (spirit) must be manifesting itself. The pre-conquest Maya were familiar with the principle that all matter is vibrating energy long before Einstein wrote his famous 1905 paper on the equivalence of matter and energy.[114]

A Native worldview requires respect for all living things, as the Lakota expression *mitakuye oya'sin* ("all my relatives," or "we are all related"), which is often declared at the end of a prayer, implies. Respect implies that our relatives have the right to exist, that they have a purpose for having been created, and that this purpose is dynamic; *i.e.*, they work out their purpose in relationship with the two-leggeds. The people live in deference to a physical and spiritual universe through a respectful relationship. They pattern their lives according to natural law and follow the natural cycles. The earth itself is a system of cycles from the smallest subsystems to the largest, and life consists of continuous cycles of renewal.

Nothing is at rest. Subatomic particles transform into energy, which is vibration, and back into particles. Atoms have their own frequencies and discrete energy states. They emit photons of energy. The sun's rays are electromagnetic energy. An aura of energy surrounds every human body. The earth and the moon move cyclically in orbits. Solar systems and galaxies are also in motion relative to other bodies in space. Everything is in motion and cycles are everywhere evident. These also belong to the Indigenous world where they have their own Indigenous ways of understanding.

HISTORICAL OVERVIEW OF WESTERN SCIENCE

W estern civilization traces its origin to the 6th and 5th centuries BC in ancient Greece, particularly the cities of Ionia along the coast of the Aegean Sea where relevant philosophical developments seem to have taken place first. Ancient Ionia is located in Asia Minor in present-day Turkey near the historical city of Smyrna.[64]

Greek tradition tells us that the cities of Ionia were founded by colonists from the other side of the Aegean Sea toward the west. Other cities established in that region by Greek colonists included Ephesus, Miletus, and Pergamum. The significant poetry and philosophy that flourished there was the result of contact with the art, religion, trade, and learning activity from other cultures with whom the Greeks came in contact.

This chapter focuses on people, philosophies, institutions, governments, concepts, cultural values, and events in the history of the West that relate to science and how scientific ideas

---

[64] Smyrna is one of the seven cities mentioned in the first chapter of the book of Revelation in the New Testament.

endured, developed, or were censored.[65] My primary source is Lindberg's *The Beginnings of Western Science*.[115] We will attempt to discern significant changes that may have occurred during the historical development of Western science—those that constitute not only technological innovations, scientific discoveries, and the introduction of new theories and mathematical applications but also modes of thought, moral values, and prevailing philosophies at different times over several centuries. In so doing, differences and similarities between Western and Indigenous thought should become evident.

Presocratic Masters of Wisdom

Questions about the ultimate constituent of matter, the nature of knowledge and change, and the duality of matter and mind were some of the issues that threatened the unity of knowledge among Greek philosophers in the 5th century BC.

Socrates (c. 469-399 BC) is known though the written accounts of his disciples and other philosophers. He was Plato's teacher. The early Greek masters from Ionia who preceded Socrates—the "Presocratics," as they are now called—particularly Parmenides, Empedocles, and Pythagoras, who flourished approximately between 600 and 430 BC, embraced a cosmology that was spiritual and holistic. This aspect about them, involving the prevalence of mysticism and spirituality and the conception of a unified world that made no distinction between animate and inanimate or between spirit and matter, is unfortunately not well known. In Fritjof Capra's words (*The Tao of Physics*): "In fact, they did not even have a word for matter,

---

[65] Although a variety of sources were consulted, Lindberg's *The Beginnings of Western Science* was the primary source on the history of science; however, not every instance in which a reference to the book was made has been cited.

since they saw all forms of existence as manifestations of the *physis*, or nature, endowed with life and spirituality."[116]

But in the 5th century BC, ominous questions arose among the Greeks which threatened this unity of knowledge and spirit. Some of them abandoned their spiritual heritage, leading eventually to a great loss for Western civilization. Questions about reality and material causation surfaced: What do things consist of? What is the ultimate constituent of the cosmos? Is change possible and, if so, how does it occur? What is the nature of knowledge and how is it obtained? Can our senses be trusted to reveal what the intellect tells us? On the one hand, the senses reveal color, taste, and so on; on the other, reason tells us that only atoms and the Void (empty space) exist. If the senses are not trustworthy, we must abandon them as a guide to truth. If the fundamental building blocks of which matter consists are passive, how do motion and other forms of change occur? Some of the Presocratics claimed that things behave according to their inherent natures, which alone could account for order and regularity in the cosmos. But if so, then order arises from within, ruling out any outside causes.

We must ask how it came about that the spirituality of the Presocratics was apparently lost and forgotten while the teachings of other Greeks, particularly Plato and Aristotle, not only survived but were enthusiastically conveyed across the centuries. Part of the answer is that Aristotle took an interest in the material realm and developed a system of knowledge with a characteristic emphasis on empiricism, as reflected in the dictum, "If it can't be measured, it doesn't exist." He would eventually create an organized system for natural science, including its practical, poetic, and theoretical aspects, that was transmitted to the entire Western World. His works were extensive, including physics, metaphysics, poetry, theater, music, logic, rhetoric, politics, government, ethics, biology and zoology.

But there is a darker side to this story than Aristotle's interest in secular science: Kingsley, who has examined the surviving fragments of Presocratic writings, discovered that the Presocratics, particularly Empedocles, Parmenides, and Pythagoras, were masters of wisdom who held a comprehensive and holistic view of the Universe and gave the West the foundational principles upon which to build a civilization, which clearly contradicts the prevailing, now standard, view of later Greeks, notably Plato and Aristotle, who wrote that the Presocratics (the true founding fathers, says Kingsley) merit no more credit than being recognized as the purveyors of primitive scientific thought.[117]

Other Presocratics

The split mentioned above involved a new kind of discourse affecting the way Greeks thought, talked, and believed. Those from Miletus saw no distinction between what is animate and what is inanimate; *i.e.,* between spirit and matter, for there was no word for "matter" because *physis,* or nature, which was endowed with life and spirituality, manifested itself in all physical forms of existence. They began to speculate, to ask questions about evidence, and to debate. A new tradition of inquiry was born involving critical assessments and purely naturalistic explanations that excluded the divine (the gods) and the spirit. A new way of thinking, questioning, and investigating emerged and became known as "natural philosophy." It preceded "natural science" and was the precursor to modern science. According to Lindberg, Aristotle referred to the natural philosophers as *physikoi,* a term that the earlier Presocratics also used with a somewhat different meaning.

The Condor and the Eagle

Heraclitus of Ephesus believed in a world of perpetual change, of eternal "becoming," without beginning or end. He claimed that the world is balanced and stability is maintained even when change occurs because of the unity of cyclic interplay between opposing forces. But *Parmenides* and other philosophers of Elea strongly opposed Heraclitus, claiming that change is logically impossible; something cannot spring from nothing, from nonexistence into existence. This thought from the Eleatic school continued, creating a conflict that was never reconciled. Eventually there arose the atomistic concept that the cosmos consists of an infinity of tiny atoms (Greek *atomos*: "that which cannot be cut into smaller pieces"), too small to be seen, moving randomly in an infinite Void (empty space). The atomists thus created a dualism between matter and spirit, the natural and the divine, respectively.

Plato (428 – 348)

In this and the next section, I have endeavored to summarize Plato's and Aristotle's views on the questions that spawned divisions among the Presocratics and led to the split in the 5th century. These questions relate to the following concepts: reality, change, knowledge, and possibly time.

Plato, a disciple of Socrates and later Aristotle's mentor, founded and conducted a private academy of philosophy about six miles outside the city walls of Athens. Owing to their intellectual endeavors, these three philosophers are credited as the founders of Western philosophy. But, as I've already noted, this view is now questionable.

Plato's legacy is nonetheless important because of his influence on the development of Western philosophy; he and Aristotle began a tradition that was propagated through the Middle Ages and later on to modern times. Plato's *Timaeus* is primarily an account of how the Demiurge—the "divine craftsman," a term that corresponds to Creator—formed the cosmos. Plato, like Aristotle, was critical of the Presocratics, claiming that they deprived the world of divinity and purpose.[118] He reasoned that the Presocratics could not offer a *causal* account of the cosmos because of their belief that things behave according to their inherent natures—a belief which implied that order in the Universe is intrinsic and with no outside cause.

On the question of *reality*, there are two realms in Plato's cosmology: The realm of *forms* is the primary reality, perfect, eternal and unchanging, for they are dependent on nothing else for their existence; the *material* realm, which consists of imperfect replicas of the forms. The cosmos and everything in it are, therefore, replicas of the forms. The Demiurge, or Creator,[66] is a rational, divine, and benevolent craftsman—a mathematician who constructed the cosmos on geometrical principles and imposed order according to a rational plan. It was not creation out of nothing; rather, the craftsman constructed the cosmos out of a primitive chaos that was filled with unformed material. Each physical object consists of particles, where each particle is a geometrical solid consisting of triangles as the indivisible "atoms."

Plato accepted Empedocles' four elements (which Empedocles called "the roots of all things")—fire, air, water, and earth—and reduced them to geometrical solids, of which there are exactly five. This scheme was a step toward mathematization of nature. The cosmos is spherical because it is the most beautiful and perfect shape.

On whether *change* is possible, Plato's response was affirmative; he assigned those things that are subject to change

---

[66] The capitalization is mine, based on the capitalization of Demiurge, a word derived from the Greek word for craftsman, or public worker; however, the Demiurge is divine.

to the material realm and eternal, unchanging things to the realm of forms. Change and diversity occur by combining the primary elements to produce variety in the material realm.

The knowledge of forms can be obtained only through the exercise of reason unaided by the senses. Knowledge in the material realm is grasped by opinion or with the senses; matter "comes to be and passes away," like the cosmos. Plato also said that "we are *souls imprisoned* in bodies" that need to escape but our *senses* are chains that tie us down.

Plato's conception of *time*, described in the *Timeaus*, is not clear. He conceived of the stars, moon, and sun as moving in their orbits to serve as markers of time and of time itself coming into being with these celestial movements. Because the movements were of ceaseless duration, they provided a moving image of eternity. Aristotle rejected this cosmology not only because the universe required a beginning (since in his view, it was created) but also because time itself needed a beginning, a concept that made no sense to him. Some scholars claim that Plato's creation story is not meant to be read literally but metaphorically.

Time seems to be a kind of celestial clockwork or motion; time *is* motion. When the Demiurge created the universe, he also created time—a kind of motion but not the measure of motion. For Plato, time *is* celestial motion.

Here is an interesting coincidence: Plato defined the "perfect year" as the return of the celestial bodies (planets) and the diurnal rotation of the fixed stars to their original positions. The term "great year," on the other hand, is defined by NASA as the period of one complete cycle of the equinoxes around the ecliptic, about 25,800 years, due to the earth's precessional motion (which was unknown to Plato), also known as the "platonic year." The last precessional cycle ended on the winter solstice, December 21, 2012.[67]

Aristotle (384 – 322)

Aristotle was a student of Plato and a member of his academy for twenty years before starting his own school of philosophy. His most powerful influence, zoology, was felt all the way to the Renaissance and beyond due to the explanatory power of his philosophical and scientific systems. One of Aristotle's students was Alexander the Great.

Aristotle was thoroughly familiar with Plato's teachings, having been his student for twenty years and a member of his academy, but he repudiated the inferior and dependent status Plato had given to the material world. In Aristotle's cosmology, which is radically different from Plato's, material objects were the *primary* realities consisting of inseparable *form* and *matter*. In his concept of matter and form, the form of an object was contained within its structure instead of an abstract characteristic; thus, the form could be perceived by using one's senses. Like Plato, he accepted the four basic elements that comprise the world: earth, water, air, and fire (which are actually Empedocles' "roots of all things"). Aristotle maintained that properties cannot exist independently; a property has to be the property *of* something. Form consists of properties, such as color and weight, which identify the object; matter is the subject or substratum for the form. For living things, the soul is the form of an organism; it has the responsibility for vital characteristics that living things possess. He did not accept Plato's regular solids and, instead, postulated sensible qualities as the ultimate building blocks.

---

[67] This date also marks the end of the Mayan Long Count calendar, which began in 3114 BC.

He argued against atomism, believing that matter was continuous; there is no void (empty) space; the universe is full. One of the problems that atomists had to address was explaining cohesion (for example, how does a rock stay in one piece); also, where does coldness come from and where do characteristics of living things come from?

He postulated that knowledge begins with sense experience; it is empirical. Repeated sense experiences produce memory, which in turn generates intuition or insight. This gives an experienced investigator the ability to discern universal features, or properties, making it possible to use "deductive demonstration" as a premise. For example, an experienced dog breeder can come to know the essential characteristics of a dog without which an animal cannot be considered a dog.

The world consists of *place* as well as *space*. There is a natural straight-line motion with each object; motion in any direction other than the natural one is due to coercion from an external force. Motion in the celestial sphere (see below) is continuous, uniform, and circular. The "fixed" stars move with perfect uniformity. Aristotle postulated his famous theory of change consisting of four causes: *formal* cause is the form or shape the object takes ; *material* cause is what an object consists of; *efficient* cause is the force that affects the object; *final* cause is the object's purpose. Consider a statue as an example. The formal cause is its shape; the material cause is the marble of which it's made; the efficient cause is the sculptor; the final cause is the statue's purpose.

As to the cause of motion, the Prime Mover, a living deity representing the highest good, is the *final* cause of movement in the heavens. Aristotle considered the eternal universe to be a great sphere, consisting of concentric shells. The celestial region, or outer shell, is filled with an incorruptible *quintessence* (an unchanging fifth element, the aether), which comprises the heavens. There are several celestial spheres, one for each planet, and each is filled with quintessence. Each sphere is moved by its own unmoved divine mover, rather than a single Prime Mover for the whole cosmos.

Aristotle's concept of time was intimately related to change and movement. Change cannot occur without time; thus, it can only be understood in terms of change and not vice versa. Time is defined as something that can be counted; it emerges when the mind (soul) is aware of two instants of time.

It is my humble opinion that even the great Aristotle was unable to explicate time so long as he assumed it to exist as a real entity. In my own view, following Einstein's judgment, time is a useful tool but does not actually exist.[68]

The Middle Ages (AD 500 – 1450)

One of the most important facts about this period in history was the survival of Greek science; had it not been disseminated and taught in and through the Middle Ages, it would have been completely lost to posterity. This period also marked the collapse of the Western Roman Empire (at the beginning of the period), the technological and agricultural inventions achieved during the middle of the period, and the poverty, war, and the Black Death plague near the end of the period.

The classical tradition and natural philosophy gradually declined in the Roman Empire in the Latin West, whereas the consequences in the East were less severe as the old Roman Empire became separated, giving rise to the Byzantine Empire. By 524, the Latin West had been largely cut off from original Greek science and natural philosophy; it experienced a steep decline but still had Plato's Timaeus, some of Aristotle's logical works, and other bits and pieces.

---

[68] If this statement is confusing, see my comments about time and time interval in the section "Space and Time" in chapter 7.

The Condor and the Eagle

One notable figure in the history of Roman science was Pliny the Elder (AD 23 – 79), the author of *Natural History*, an encyclopedic work that covers almost the entire field of ancient knowledge. His work was highly referenced and translated all through the Middle Ages. He was also a naturalist or natural philosopher and an important naval and military commander. His work was used for reference over the following centuries by countless scholars, especially in medicine, plants and plant products, agriculture, architecture, sculpture, geology and mineralogy.[119]

The Roman Empire began with the appointment of Julius Caesar in 44 BC after existing as a republic for 500 years. In AD 285, Emperor Diocletian divided the Roman Empire's administration into the Latin West and the Greek East and appointed a separate emperor for each, one in Rome and the other in Constantinople. The Western Roman Empire fell to the German Barbarians in AD 476, marking the start of the Middle Ages, or the Medieval Period. The Eastern Roman Empire, a.k.a the Byzantine Empire, lasted more than one thousand years until AD 1453.

The Middle Ages was a time of Greek influence and dissemination of natural philosophy, also known as Greek science. *No new scientific theories were introduced during this period*; nonetheless, significant activity during this time in history was in support of science. Important scientific works, many of which were translated into other languages, were preserved and science education was substantial.

The dissemination of Greek science around Europe and Western and Central Asia became possible through the translation of Greek into other languages. Greek manuscripts were copied and Greek science was taught in schools and eventually in the universities. Greek culture was assimilated slowly but continued for more than one thousand years and benefited even the aristocracies of Western and Central Asia and Islam. The Islamic world took advantage of paper-making technology (imported from China) to copy manuscripts. Hippocratic writings, consisting of the medical treatises of Hippocrates (the 5th century BC Greek physician), and other treatises, were translated from Arabic to Latin, serving as the foundation from which the West built its medical literature.

During the period AD 300-500, the Assyrians translated Greek scientific, religious, and philosophical writings into Arabic, Syriac, and Assyrian (Syriac-Aramaic). When Assyrian versions were later translated into Arabic and brought by the Moors to Spain and translated into Latin, they helped spark the European Renaissance—a new era of scientific inquiry and appreciation of ancient Greek and Roman civilizations made possible by Moorish advances in mathematics, astronomy, art, and agriculture.

In 711 a Muslim army landed in Gibraltar (Spain) with around 10,000 men, mostly Berbers (Indigenous North Africans), and in seven years conquered the Iberian peninsula, becoming one of the great Muslim civilizations with Southern Spain or Andulusia (*Al-Andalus*) as the heartland of Muslim rule, lasting until 1492 when Granada was conquered by the kingdom of Ferdinand and Isabella. From 711 to 1492, Islamic Spain, populated by Christians, Jews, and Muslims, was under varying degrees of Moorish (Muslim) rule but had a long history of living in harmony until 1492 when the Jews were expelled.

In 1492 Granada was captured by the Moors; in that year Columbus's first journey to the so-called "new world" and the seeds of overseas empire were planted. Printing developed in the middle of the 15th century, greatly expanding the availability of books and bolstering literacy, thus helping to fuel cultural changes and advances in the sciences and humanities.

In education, Europe saw the first universities beginning circa 1200 when the strongest schools were upgraded to universities, the first ones of which were located in Bologna, Italy (1150), Paris (c. 1200), and Oxford (1220).[120] Aristotelian natural philosophy became central to university curricula in mathematical sciences. Aristotle's and Plato's works became permanent contributions to Western thought.

The strongest schools endeavored to recover and master the Latin classics (or Greek classics that were available in ancient Latin translations), surpassing anything seen in the early Middle Ages. Urban schools saw a marked "rationalist" turn—an attempt to apply intellect and reason to many areas of human enterprise. Rationalism—the attempt to explain everything with the intellect and reasoning powers—became strong and there was a persistent effort to apply Aristotelian logic to theological problems. For example, theological argumentation replaced biblical authority to prove the existence of God.

In natural philosophy, Isidore of Seville and Bede the Venerable stand out, both of whom exemplify how science was preserved and promoted during the Middle Ages. Although they typically did not contribute to scientific discovery, without them we would have no understanding of how we got where we are today. Isidore, an outstanding scholar in biblical studies, theology, liturgy, and history, wrote two books: *On the Nature of Things* and *Etymologies*, the latter of which covers seven liberal arts, medicine, law, timekeeping and the calendar, theology, anthropology, geography, cosmology, mineralogy, and agriculture. He believed the cosmos was geocentric, composed of four elements, in a spherical earth, and could give an account of the celestial sphere, the seasons, the nature and size of the sun and moon, and the cause of eclipses. Bede, a priest and an English monk, probably the most accomplished scholar of the 8th century, is best known for his *Ecclesiastical History of the English*

People, On the Nature of Things (based on Pliny the Elder and Isidore), and two textbooks on timekeeping and the calendar.

In the 12th and 13th centuries many of the texts containing scientific achievements of both Islamic and Greek origin came into the possession of Latin Europe in the West, where they were enthusiastically received, and renewal of cultural transmission began. By the second half of the 13th century, Aristotle's works on metaphysics, cosmology, meteorology, psychology, and natural history became compulsory subjects in the universities. The sudden influx of Greek and Arabic learning through the translations of the 12th century made it possible for the universities to have a common curriculum. Instructors (called "masters") not only enjoyed a high degree of freedom but also had common credentials that allowed them to teach in *any* school.

By the year 1000 *feudalism* had become the dominant social, economic and political system in the West, began to decline in the 13th century, and lasted into the 15th century just before the Renaissance. The supreme monarch gave land to nobles and demanded their loyalty; the nobles gave land to vassals, who served as knights to defend their monarch or noble; the peasants or serfs at the bottom were under the vassals' control.

Regarding the Church's[69] influence on science, it can be said that the Church "motivated" scientific inquiry but it did so in a strange way by applying Plato's cosmogony, and his forms and essences, to physics and cosmology for the purpose of elucidating the Genesis account of creation. Science functioned as the Church's handmaiden, which explains why Plato's *Timaeus* seemed more popular than Aristotle's works. The Church discouraged investigations by means of empirical

---

[69] This particular reference is to the Church of England during the Middle Ages, founded in the 6th century AD. Later references to "the Church" or "the churches" generally point to institutional Christianity.

methods because it had adopted the Greek view of nature as eternal and self-existent rather than created; according to this view, a living organism was imbued with attributes of divinity.

The Church's rejection of empiricism during the Middle Ages on the basis of a mystical or spiritual view of nature, which it had adopted from the Greeks, may be may be due to several reasons, but it is the *opposite* of what I believe to be a successful approach today for being accepted by the scientific establishment, although there is a difference between institutional and individual goals and purposes. The Church in that era may have been resistant to questions that could disprove or jeopardize its doctrinal position. Or, it may have been asserting its authority over all affairs, sacred and secular. On the other hand, it did not have the advantage of modern physics concepts which today are friendlier to a holistic cosmology.

The imbalanced relationship between the Church and science notwithstanding, the Church's belief in the existence of *order in the cosmos* did play an important role in encouraging scientific inquiry. One of the criticisms against the Church was its anti-intellectual stance but, ultimately, Christianity became the major patron of education in the Latin West because Christians recognized that encouraging people to read the Bible was a way of improving literacy. Thus the Church reached out to the upper classes, including the educated, by promoting the Bible as literature.

The Church was one of the most powerful, if not *the* most powerful, institution in Medieval Europe. It helped launch many of the great universities of Europe (see below) and sponsored art, music and architecture. But the Church's extreme power and wealth spawned a reform movement that led to the founding of the Dominican and Franciscan Orders, which addressed the problem of poverty and emphasized spirituality.

Large Gothic cathedrals and Gothic architecture proliferated throughout Europe which, as best as can be ascertained, were an expression of humility before a majestic and powerful God; they were intended to connect (or re-connect) people to heaven and the divine.

Note that the style of architecture which was created to counter the image of the Church as a powerful and wealthy institution reflected a certain worldview. Western culture's concept of humility, at least at that time, was to build majestic and costly structures as a symbol of God's majesty in contrast to human frailty. If this image is also a response to human poverty, it seems either to equate spiritual poverty with physical poverty or to focus on the spiritual and neglect the physical. How does it help people who are poor and hungry, or more to the point, how is it consistent with Jesus' admonition to help the outcasts in society: "For I was hungry and you gave me something to eat, I was thirsty and you gave me something to drink, I was a stranger and you invited me in, I needed clothes and you clothed me, I was sick and you looked after me, I was in prison and you came to visit me. ... whatever you did for one of the least of these brothers of mine, you did for me."?[121]

Islamic Science[122]

The Islamic Empire was in power from AD 636 until AD 1258. The Byzantine Empire had risen to power three centuries earlier and initially provided the medieval Islamic world with Ancient and early Medieval Greek texts on astronomy, mathematics and philosophy for translation into Arabic, since the Empire was the leading center of scientific scholarship in the region in the early Middle Ages. Byzantine science was essentially classical science from ancient Greece and ancient Rome, so it is closely connected to ancient Greek philosophy and metaphysics. During the Middle Ages, there was frequently an exchange of works between Byzantine and Islamic science. Later as Arab and other

Muslim cultures became centers of scientific knowledge, Byzantine scientists translated books on Islamic astronomy, mathematics and science into Medieval Greek, written mostly by Persian scientists.

Islamic science benefited greatly from the fact that Greek science was available in the form of books rather than fragments requiring translation and compilation. Its greatest achievements were in mathematics and astronomy, which originated in Baghdad. Islamic astronomy flourished well into the 16th century, even after the birth of European science. The achievements of Islamic science included the following: First, algebra (from *al-jabr*, which means "to restore"; *i.e.*, to transfer quantities across the equals sign), which originally had no algebraic equations or symbols; these resulted from using Euclidean geometry to solve problems. Second, geometry based on *Euclid's elements*, a set of thirteen books containing definitions, postulates (axioms), propositions (theorems and constructions), and mathematical proofs of the propositions). Third, Islamic trigonometry was based on Ptolemy's *Almagest*, one of the major Greek works that was utilized and translated all through the Middle Ages. Ptolemy, from the first and second centuries AD, was a Roman mathematician, astronomer, geographer and astrologer, who authored several scientific treatises, of which the *Almagest* is one. The fourth achievement was mathematical astronomy (the greatest achievement), which demonstrates the relationship between Islamic and Greek science. Islamic astronomical activity consisted of observation, planetary modeling (including the use of mathematics to make predictions), and observational programs. Other contributions were geometrical optics and medical science, including the building of hospitals.

The 16th and 17th Centuries

Lindberg (*The Beginnings of Western Science*) attributes the rise of modern science in the 16th century to a change from a *qualitative* to *quantitative* mode of thought under which science was conducted; *i.e.*, from *why* things happen to *how* they happen. This new impetus was characterized by an obsession with detail, experimentation, objectivity, and the synthesis of science and technology. The manuscripts of Aristotle, Plato, and the genius of earlier Greeks provided the influence for this new trend.

Modern science was no longer like Greek science, although the latter had played an essential role. Greek cosmology emphasized purpose (final causes), a complete theory of *why* things happen but not *how* things happen (efficient causes), as exemplified by Galileo (1564-1642). In contrast, modern science wants to know facts; it seeks to know and to analyze the minute facts of nature.

Copernicus, Newton, and Vesalius were key figures during the 16th century. Copernicus (1473-1543) applied mathematics to his analysis to disprove Ptolemy's theory that the earth is the center of the Universe (with the sun revolving around it instead of conversely), which had been accepted for over one thousand years. Isaac Newton's famous *Principia Mathemática* (Mathematical Principles of Natural Philosophy), containing the laws of motion, universal gravitation, as well as the derivation of Kepler's laws of planetary motion, was published in 1687. In it, he synthesized the work of Copernicus and other prominent figures—Tycho Brahe, Johanne Kepler, and Galileo Galilei. Vesalius wrote the first anatomically accurate medical textbook, *De Humani Corporis Fabrica* (1543), which was complete with precise illustrations, and much more.

Newton banished the belief that motion on the earth and motion in the celestial sphere obey different physical laws (known as the Aristotelian terrestrial/celestial dichotomy) which had dominated intellectual thought for nearly two thousand years.

My own focus of study in graduate school was theoretical physics, a.k.a. mathematical physics, which began in the 16th century as the synthesis of two previously separate disciplines that separated mathematicians from natural philosophers (or physicists) and represented a fundamental change in the investigation of nature. Galileo, for example, experimented with gravity by rolling a ball on an inclined plane, calculated its acceleration, and determined that the distance traveled is proportional to the time squared (t^2). In other words, he compared experiment with theory—a quantitative approach. Others who applied mathematics to the physical world were Copernicus, Galileo, Kepler, and Aristotle. As these examples demonstrate, the success of Western science was in large part due to the following: the quantification of empirical observational data, expressed mathematically, to prove the correctness of an idea about nature's behavior, and the skill in classifying only those things that can be observed (by the senses) and measured.

A major conclusion that was drawn implicitly as the result of repetitive experimentation was that *there is order in nature*, that nature can be trusted to behave the same way under the same conditions, giving researchers the confidence that, by repeating an experiment several times and obtaining the same results, the behavior of a phenomenon being studied can be determined. From a Western perspective, this approach produces a step in the advancement of knowledge, which is a major goal of science.

From my perspective, the order in nature, which was recognized long ago, is only one aspect of a relationship, and our relationship with nature is what's important. Knowledge comes from the experience but it may be unique to the individual. And, of course, a Native individual would do much more than simply watch a ball roll down an inclined plane.

According to the Encyclopedia Britannica, Isaac Newton immersed himself in the works of Aristotle, which could of course encompass any number of topics: physics, metaphysics, poetry, theater, music, logic, rhetoric, linguistics, politics, government, ethics, biology, and zoology. The natural philosophers, except Aristotle and Archimedes, practiced science without using analytical techniques. Had the Greek classics, which included Aristotle's and Plato's works, not been preserved during the Middle Ages, they would have been unavailable to Newton and everyone else living in the 16th century and later.

Alfred North Whitehead called the 17th century the "century of genius," citing three main factors: the rise of mathematics, the belief in a detailed order of nature, and the abandonment of the metaphysical analysis of the Middle Ages in favor of empirical facts.[123] Galileo is important because (a) he applied mathematics to motion and (b) his observations of the moon, the planets, and the Milky Way (using the telescope he built—the first one of astronomical quality) vindicated Copernicus. Despite imprisonment, he still succeeded in discrediting the Church.

The philosophy of Descartes (1596-1650) became a foundational element of Western philosophy. His famous assertion, "I think, therefore I am," has been explained in numerous ways but the simplest seems to be based, not on thinking, but on doubting: "I doubt, and because I doubt, I know I exist."

Atomism (from Greek *atomos*: "that which cannot be cut into smaller pieces"), the notion that the universe consists of lifeless bits of matter, which first emerged in the 5th century BC, returned in the 17th century and remained popular at least until the advent of quantum theory. It was further adopted by prominent figures like Galileo, René Descartes, Robert Boyle, Isaac Newton, and others. The philosophy of *mechanism* became dominant, replacing Aristotle's natural philosophy of purpose and order in nature, but physicists could not explain light by applying Newton's laws of motion.

The 17th century's answer to the Ionian (Ionia in ancient Greece) question, "What is the world made of?" is that the great forces of nature were entirely determined by the configurations of material objects or particles. The famous *mechanistic theory* of nature, or materialistic mechanism, reigned supreme like an orthodox creed, even into the 20th century, and physicists apparently had lost interest in philosophy. At the close of the 17th century, science and philosophy split into mutually exclusive domains: science focused on the material world and philosophy focused on the mind, reminiscent of the 5th century BC when a split occurred among Greek philosophers.

The 18th and 19th Centuries

In his overview of 17th and 18th century science, Whitehead laments the materialistic mechanism that characterized that era. It paid no attention to the concept of a living organism (as opposed to the concept of dead matter), including the human being; authors of literature chose not to comment on the philosophical aspect of science; they simply ignored it. As an example, a verse in one of Tennyson's poems—"The stars," she whispers, "blindly run"—implies that the physical world is simply there like the molecules in human beings. Scientists portrayed a world that functions without meaning or purpose. In contrast, Whitehead's "philosophy of organism," or "organic philosophy" reflects his view that *reality consists of interrelated and mutually dependent parts* which are involved in sustaining vital processes.

During the era known as the Age of Reason,[70] the philosophy and cultural life in the 18th century represented a departure from the religious authority and intolerance of the Middle Ages and the advocacy of reason as the primary source of authority in preference to tradition, faith and revelation. Whitehead referred to it as a complete antithesis to the Middle Ages.[124] The notion that all of the processes of nature can be explained mechanistically, rejecting the supernatural, became a dogma of science. The movement began in Europe and spread to the American colonies where the emphasis on liberty, democracy, republicanism and religious tolerance culminated in the drafting of the Declaration of Independence. Two famous deists were Thomas Paine and Thomas Jefferson, whose short Jefferson "Bible" had removed all aspects of the supernatural. Deists (from Latin *deus* - God) embraced a Newtonian worldview, believing that everything in the universe, even God, must obey the laws of nature.

Lagrange (1736 –1813) re-formulated classical mechanics by combining conservation of momentum with conservation of energy. His *Mécanique Analytique*, published in 1788—a century after Newton's *Principia*—was a triumph in mathematical physics. Electromagnetism was thought to require a *continuous* medium, something that can undulate, and because of its wave nature, light was known to consist of electromagnetic radiation, so the hypothesis of a luminiferous ether as a propagating medium covering the whole earth was still deeply ingrained.[71] James Clerk Maxwell's publication of *Electricity and Magnetism* in 1873 was another milestone in physics because he demonstrated that electricity, magnetism, and light are all

---

[70] Also known as the Enlightenment or the Age of Enlightenment. The *Age of Reason* is also the title of a pamphlet by Thomas Paine, who articulated the deists' position.

[71] The Michelson-Morley experiment was not conducted until 1887.

The Condor and the Eagle

manifestations of this same phenomenon he called the *electromagnetic field*. These three books alone were amazing triumphs; however, other mathematical physicists— Maupertuis, Clairaut, D'Alembert, Lagrange, Laplace, Fourier, Carnot, Gauss, and Riemann—also made important contributions during the 18th and 19th centuries.

The incredible triumphs of 18th century science notwithstanding, commentary about this period would be incomplete without mentioning the Romantic Movement of the late 18th and early 19th centuries, which began in Britain and subsequently spread throughout Europe, North America, and eventually the world. The Romantic era came about in response to several profound changes affecting the way in which society was accustomed to live and work: People who customarily worked in homes were replaced by mass-producing machines in factories; the population doubled; nationalism, by which individuals who belong to a territory and embrace a common language and history, was born; they were now patriots connected to a state system and eligible for military service. It was no longer the same society socially, culturally, or economically. They became a society that relied on activities which could satisfy emotional human needs while, on the other hand, trusting in the advances of science and technology."[125]

Romanticism was partly a reaction to the rationalism and intellectualism of Enlightenment culture as well as a revolt against the impersonal treatment resulting from a new era created by the Industrial Revolution. Romanticists expressed their feelings and belief in the inherent goodness of people and nature through the visual arts, music, and literature. Feeling surpassed reason; being close to nature, which society had not yet managed to corrupt, meant being close to God. In fact, for many, God and the Universe were one and the same. The philosophy of *transcendentalism*, which was a reaction against the rigidity of society's institutions including organized religion and political parties, developed in the Eastern portion of the United States from the influences of Harvard Divinity School and the Unitarian church. Romanticism also attracted those who specifically opposed the rigid and fatalistic character of Calvinism, whose doctrine of predestination and related doctrines teach that God's elect are those chosen for salvation, even before birth; *i.e.,* they were "predestined," which implies that individuals are not truly free to chose their own eternal destiny.

The 20th Century to the Present

The Enlightenment of the 17th and 18th centuries began in Europe and spread to the American colonies, promoting the advancement of knowledge through science and reason. It was an intellectual movement that relied on empiricism to verify truth while reacting against the religious intolerance and other abuses that occurred during the Middle Ages. Several of the abuses are, of course, mentioned specifically in the U.S. Declaration of Independence. The scientific developments of the 17th through the 19th centuries contributed to the development of modern thought which inhibited a humanistic worldview. By the beginning of the 20th century, the mood in science was one of optimism; the Universe was perceived to be almost completely knowable and, owing to the human ability to make rational decisions, the world could only expect a better future.

The attitude among scientists began to change after the quantum behavior of energy was discovered; compared to 17th century science, a "new" science was clearly in view in the 20th century. As already stated, due to the strange behavior of the quantum and Einstein's achievements in the special and general relativity theories, the question about the nature of reality, which was debated by the Greeks during the 5th century BC, re-emerged with a new vigor in the 20th century and, as we now know, ours is not a world of inert atoms and mechanical causation but one that is infinitely more complex than previously imagined.

According to Alfred North Whitehead, whose insights are relevant in view of his status as a contemporary and renowned philosopher and mathematician of the early 20th century, the dominating issue of the 17th through the 19th centuries was that scientific concepts were based on materialistic doctrines— evolution, energy, and molecular theories—instead of orthodox materialism, which, as he also noted in his lectures on *Science and the Modern World* (1925), no one had pointed out.[126] It was an outlook that presupposed the ultimate "fact" of irreducible brute matter, which was characterized as senseless, valueless, and purposeless.

The issue of materialistic thinking and its ramifications has not disappeared and may never entirely disappear; however, the current noticeable trend toward holism in science may promote more wholesome attitudes due to influences driven by compassion, intuition, spirituality, and the desire for balance. Meanwhile, another issue that scholars outside of physics have vocalized is the prominence of a haughty attitude within the scientific establishment. This lament is not surprising because the alleged attitude is commensurate with the achievement of hegemonic status on the part of the eagle, as foretold by the Prophecy of the Eagle and the Condor, according to a common interpretation of this prophecy. The condor and the eagle must use the opportunity now available to fly together as before, wing to wing, and achieve a balance of gifts between the north and the south. It takes a small leap of logic to connect the eagle with Western science.

As the world entered into the 20th century, even before the advent of quantum theory, some philosophical shifting, mostly related to changes in worldview, had already occurred which had an effect on scientific concepts. The first documented change was the rise of atomistic views among the Greeks in the 5th century BC, contradicting the original view that matter and spirit are united and that nature is alive. The atomists, you will recall, theorized that the natural world consists of an infinity of tiny, indestructible, inert, and passive "atoms" (indivisible particles, not a reference to the modern atom) too small to be seen, moving randomly through the *Void* (infinite, empty space). Atomism is a dualistic philosophy which holds that the natural and divine orders are separate entities. *Scientific materialism*, which holds that the Universe consists only of matter (*i.e.*, matter is the only substance that exists) and that all phenomena result from material interactions, was the dominant philosophy from the 17th through the 19th centuries.

It should not be difficult to see why this philosophy became meaningless and worthless soon after the birth of modern physics: In 1905, Einstein proved that matter and energy are equivalent; thus, to say that only matter exists in the Universe is equivalent to saying that only *matter or energy* exists. This implies that everything in the Universe was derived from matter or energy as the ultimate constituent (or substance).[72]

Did matter-energy equivalence eliminate the "evils" of scientific materialism, since *there is no such thing as a purely material object* and the matter that constitutes all material objects and particles is equivalent to energy? Isn't the affirmation of spirit what irks some scientists? Until it can be proved otherwise, energy and spirit are equally compatible with the laws of physics.[73] Science cannot satisfy the soul in the absence of spirit; it cannot address and fill the deep longings of the

---

[72] It has also been suggested that mind is the ultimate constituent. See my comments in the Space and Time section of chapter 7.

[73] This statement about compatibility is made in the context of physics and makes no assertions about tribal beliefs regarding energy.

The Condor and the Eagle

human being. This issue may be at the root of what brought about the romantic movement in the 19th century. Without spirit, our human attributes—intellectual, emotional, physical, spiritual—are unbalanced; if one of these suffers, the others become distorted. Notice the made-for-television movies that are shown at Christmastime every year: People need the magical "Christmas spirit." Many people become depressed at this and other special times of the year because a sensitive part of them is vulnerable and needs fulfillment. Materialism alone cannot fulfill these needs.

Are there scientists who need to heal from materialistic science? I have a friend who left his science-related profession. He states that the analytical mind blocks the spirituality of wholeness; that scientific materialism is potentially harming to the human spirit; that we (people in science) often internalize the scientific method in a manner never intended by excluding the spiritual from our work.

To summarize the shifting of philosophical winds: the first one was from the unity of nature to a dualistic view of matter and spirit (atomism became a new philosophy); then the second one was to scientific materialism, to reductionism and atomism (when it returned); and finally to holism in the current era of theoretical physics. The first shift was driven by human speculation, the second by human reason, and in the last shift, nature responded to a human inquiry in accordance with a prophetic timeline.

Reductionism is the theory or philosophy that the objects or phenomena which function at one level of a physical system can be explained in terms of those at some other (typically reduced) level. Like atomism, it is untenable on the basis of modern physics but it is interesting that conference speakers still argue against reductionism when it is no longer necessary.

Opposite to reductionism is wholeness, the notion that best characterizes what we have learned about the Universe in the 20th century. Based on physics research, the Universe is an undivided whole, permeated by an energy field, and can no longer reasonably be conceived as consisting of unrelated parts but as a network of relationships that connects all things are under its influence.

Non-locality: Physicist David Bohm published *Wholeness and the Implicate Order* in 1980 and for the rest of his life continued to refine a physics cosmology based on the idea of wholeness. The title of his last book, *The Undivided Universe*, on which he was putting the finishing touches (with B. J. Hiley) at the time of his sudden death, conveys his conviction about wholeness that the book expresses mathematically for physicists to examine for themselves. He states that conviction in the preface: "In the ontological theory that we present here, this wholeness is made manifest through the notion of non-locality."

Although I introduced non-locality in Chapter 6 using an analogy to demonstrate the predictive aspect of quantum mechanics (QM), I will now describe a famous paradox that will confirm that there is a mystery associated with QM's predictive feature. In a landmark 1964 paper, Irish physicist John Bell proved a theorem which provided an experimental method for testing the correctness (or incorrectness) of QM, using what came to be called Bell's inequality. Using Bell's theorem, physicist Alain Aspect ran a series of experimental tests and in 1982 reported that the predictions of QM were correct. The non-local aspect of quantum reality, which defies realism, is a fact based on mathematical proof; it is not speculation. Non-locality is still not an observed phenomenon but the correlations between properties of quanta *are* observed and provide sufficient proof. Nor can the reality that exists between the two space-like separated points be measured. Non-locality is an inferred phenomenon, a fundamental property of the Universe.

In 1935 Einstein and two of his colleagues, Boris Polanski and Nathan Rosen, designed a hypothetical "thought experiment," known as the EPR experiment, or the EPR paradox, and wrote a joint paper to challenge quantum theory by showing its presumed inadequacies; namely, its lack of completeness because it omits an essential component. Let's examine the experiment to see what constitutes this paradox:

Suppose two particles **A** and **B** have opposite spins and it is known that the total spin is zero. Since spin is a vector quantity (*i.e.*, it has magnitude and direction), we avoid working with vectors by assuming that the particles are constrained to move in only one direction. (Besides, quantum mechanics does not permit us to measure two components simultaneously because of the Uncertainty Principle, adding another detail that we don't need to bother with in this discussion.) Suppose we measure the spin of particle **A** and obtain the value s1; we know immediately from quantum mechanics that the spin of particle **B** is –s1 (since the total spin is zero). This prediction is what troubled Einstein; he argued that, since the two particles are independent, being so far apart, particle **A** could have no effect on particle **B**. He argued that, if it does, and by measuring the spin of **A** we can obtain knowledge about the spin of **B**, then **B** must have already had the value –s1, whether or not it is a measured value. He argued that, by failing to account for these definite values of spin, there must be something incomplete in the quantum theory.

Einstein's error was assuming that the two particles were independent and had definite spin values. Entangled particles are not actually independent but it is still not known how they correlate their properties, in this case their spin values. It is the non-local feature of the Universe which Einstein could not accept. However, had he lived longer, he undoubtedly would have changed his mind in view of extensive experimentation that was conducted after his death. Although non-locality is a confirmed fact, it remains an unexplained paradox.

Specifically, the EPR paradox can be stated as follows, referring to **A** and **B** as defined above: There are two possibilities and only one of them can be true: either (1) the result of a measurement performed on particle **A** has a *non-local* effect on the physical reality of another particle **B**; or (2) quantum mechanics is incomplete in that it cannot account for some element of physical reality corresponding to **B** (*i.e.*, there must be a "hidden" variable to account for it). In general terms, either the effect on particle **B** is non-local or the result of a property which is missing from quantum theory. Extensive experimentation based on Bell's theorem has confirmed that (1) is correct.

Physicist Bohm's approach to quantum mechanics is explicitly non-local and compatible with Bell's theorem despite being called a "hidden variables" theory. He rejected the Copenhagen Interpretation and went beyond obtaining knowledge *about* quantum reality in that he formulated a way that describes the reality itself by introducing the concept that the electron has well-defined position and momentum but is also affected by an accompanying "pilot" wave. Other alternatives to the CI have been proposed. I mention these facts for the sake of completeness but I do not advocate any particular theory.

By the end of the 20th century, a new concept of nature had emerged. From a physics perspective, we have learned a great deal about the ancient Greek question on the nature of reality but it still needs further examination. As I stated in the beginning of this book, my interpretations from a physics perspective are not intended to interpret Indigenous beliefs using physics.

Chapter
13

THE WESTERN SCIENCE
PARADIGM

From time to time, Larry Merculieff, coordinator for the Bering Sea Council of Elders, is given messages, stories, and prophecies from the elders. One story speaks of "a time when the gifts of the four sacred colors will come together from the four directions and combine to create something new that has not been seen since the beginning of time." Yellow brings the gift of air from the East; black brings the gift of water from the South; red brings the gift of Earth from the West; white brings the gift of fire from the North.[127]

I have heard and believe it is obvious that the gift of fire from the white direction has been abused. This powerful gift requires the proper use of the intellect and, despite the many benefits it has brought to the world, it has also brought destruction. A people's science stems from their epistemology as well as a philosophy that encompasses their doctrines and assumptions and addresses such questions as what it means to know and how knowledge is validated. Different assumptions result in a different science but, ultimately, how the gifts are used is a matter of choice.

Native peoples embrace the whole of human experience in their traditional practices rather than limit themselves to what can be measured or gathered and interpreted through the physical senses. An essential element in a Native worldview is the reciprocal relationship between humans and the natural world, including the entire cosmos. This is not just a belief system but a way of life.

Science enjoys a separate authoritative domain in our society such that the public pays close attention to phrases like "studies show," "scientists believe," "in a scientific study," and so on that are based on fragmented studies. Conclusions from these studies are frequently changing and the media that reports them keep the public confused on various issues. Yet it should be obvious that profit plays a role in many of these studies, which need to be considered with some skepticism.

Some scientists from the Western tradition acknowledge the constraints that the scientific method imposes. In *The China Study*, nutrition researcher T. Colin Campbell tells from direct experience how large corporations profit from a confused public by using science, better referred to as "scientific reductionism," for their benefit. By not telling the public everything they know about a product and how the *whole product* will affect them, customers are induced into purchasing it based on a single ingredient. It is a mistake, he tells us, to characterize whole foods by the health effects of specific nutrients.[128] Dietary supplements are one example of reductionism, and there are many others.

The conventional classroom approach to learning is out of context with the natural world and has created the need to develop relevant curricula in Indian education that incorporate aboriginal values and ways of learning. The classroom approach feeds the mind; the traditional way works through respectful relationships with our non-human relatives living in their natural environment. This characterizes the Indian meaning of "community," a communal approach that extends to all relationships in the cosmos and acknowledges their sacredness.

After more than a century since the birth of modern physics, which shattered the mechanistic and materialistic view of the Universe, the West still has not changed its way of thinking. It is not difficult to see how the assumption of a purely material world, a world without spirit, can only lead to the concept of a dead Universe that is devoid of emotion and is incapable of intimate relationships.

The Western world has long believed that everything can be reduced to lifeless objects that are supposedly unrelated to each other. The tendency to reduce things to the smallest parts makes it necessary to categorize according to some scheme. The scheme can, of course be arbitrary, such as the taxonomy in biology (…, kingdom, phylum, class, order, etc.) instead of one built on natural relationships.

It is instructive to notice how theoretical physicists are responding to the evidence of complexity in the world. Efforts to develop a complete theory of nature that unifies all physics, also known as quantum gravity, has led to a crisis among cosmologists who have worked arduously for the past thirty years in hopes of proving M-theory, but without success. In view of the incredible complexity of nature and the cosmos, which so far defy explanation, some string theorists have accepted the *Anthropic Principle*[74] as the only alternative to the problem of being unable so far to explain the complexity of the Universe. Physicists do not deny that the Universe is surprisingly friendly to complex structures and amazingly supports intelligent life as well as systems as large as galaxies. But is acceptance of the AP *by default* considered good science?

---

[74] The Anthropic Principle is described in chapter 16.

By accepting it and giving up the search for an explanation, is a physicist deviating from science?

A crucial principle that has been widely adopted by conventional (Western) scientists is the *falsifiability* of theories— an idea proposed by philosopher Karl Popper. In order to be scientific, a theory (or proposition or hypothesis) must be posited using a statement that is falsifiable, even if it is actually true and may never be falsified. Popper said that a theory can never be proved right, but if it survives several attempts to be proved false, it is worthy of being trusted (though it may eventually be falsified).

The test of falsifiability limits science to observation and experiment. It clearly gives scientific knowledge a Western connotation and, although it can usefully be applied to the scientific method, it also limits the quest for knowledge. By the Western paradigm's own definition, scientific knowledge is only a part of knowledge, since from an Indigenous perspective, knowledge is a unified whole. The question then seems to be, not what is knowledge, but what is science. Knowledge is complete; science is what we observe and accept as valid. So, with regard to a true commitment that does not limit the quest for knowledge, the error is not that spirit and the paranormal are considered valid but that they are, and consistently have been, viewed as invalid by Western culture.

The test of falsifiability limits science to measurable phenomena, but this does not imply that whatever cannot be measured is not knowledge. The test clearly separates metaphysics from empirical science. The scientific method is not equipped to deal with the totality of human experience, including consciousness, thought, feelings, intent, values, and perceptions about reality. It can teach nothing beyond how facts are related to each other.

Chapter
14

THE PHILOSOPHY OF
MECHANISM

Physicist David Bohm argued that the philosophy of *mechanism* is an extrapolation from classical mechanics,[129] an atomist theory of nature whose premise is that the Universe can ultimately be reduced to the smallest particles, or "building blocks," of matter. This philosophy also embraces the view that Newton's laws are universally valid for everything that exists in the Universe. A correspondence thus exists between mechanism and classical physics, which is the science of mechanics based on Newton's laws of motion. Classical physics is a *mechanical* system because the laws apply to particles and other objects that move or maintain equilibrium under the influence of external forces.

The idea that nothing in the Universe lies outside of classical physics is not science; it is a philosophy. Mechanism persisted in physics for a long time and it is still held among some physicists. Owing to the connection between mechanism and physics, it is worthwhile to provide a brief history of this philosophy and how it relates to concepts from classical physics, such as inertia, acceleration, and causality. It will facilitate our understanding of how modern physics (quantum theory, relativity theory, M-theory) ushered in a new order of reality that also brings us into the domain of Indigenous metaphysics.

It was once believed, as is well known, that the earth was the center of the Universe and the planets, the sun, and other heavenly bodies traveled in uniform, circular orbits around it. This system was elaborately explained by Ptolemy in the second century AD. When Copernicus proposed a simpler view—that the earth travels around the sun— much later in the sixteenth century, he met resistance. Not only did his idea contradict Ptolemy's elaborate and complicated system of epicycles but it also contradicted the prevailing doctrinal belief, supported by the Church, that the heavens were perfect. Ptolemy's system was the only acceptable theory among ancient astronomers, while Copernicus could only say, against the weight of the Church's authority, that his idea was simpler. Copernicus was right, of course, but he didn't have evidence to support his view. In the "perfect" system of astronomy, things were naturally at rest and had to be "pushed" to cause them to move; *impetus* (Latin "force") was needed.

Copernicus could not explain why, if the earth is moving, "we don't feel it." Galileo Galilei provided the answer: simple velocity is not felt; only acceleration (change in velocity) is felt. But velocity does not change unless a force causes it to change. This is how the idea of inertia (that a body resists a change in motion) replaced the old impetus concept. Isaac Newton ultimately perfected the theory of motion with his laws that express all of the causal relationships resulting from forces applied to objects in motion or at rest.

Inertia is a part of our daily experience. When we throw a baseball or open the refrigerator door, it resists with inertia. Inertia is introduced early in elementary physics courses to prepare students for Newton's laws of motion. Every object resist change in its current state of rest or motion. However, we don't measure inertia; we measure *mass*.[75] According to Newton's Second Law, the net force (the sum of all forces) applied to a body (which has mass) causes the body to accelerate.

This introduces the notion of *causality*, or cause and effect. The belief that behind every change in the Universe--regardless of what kind of change--there is a cause, spawns other philosophical questions. Is human knowledge the result of an external cause? How is what is known related to the one who possesses the knowledge? This question pertains to an important aspect Indigenous traditional knowledge, which is that the biodiversity of a particular place (the range of organisms present in a given ecological community) is the collective result of transformations created over time by the people's own science. Lorenzo Muelas Hurtado from Movimiento Autoridades Indígenas de Colombia explains that the Indigenous peoples of Colombia do not dissociate traditional knowledge from the resource; knowledge is attached to the resource and cannot be separately cataloged.[130]

Newtonian physics is a science of causality and determinism; in principle, if we know the state of a physical system at any given time, we can precisely predict its state at a later time. Mechanism is in essence the view or attitude that nothing in the Universe exists which does not fit into the scheme of Newton's laws of motion; accordingly, all things can be reduced to quantifiable laws. More subtly, however, this philosophy` contradicts the concept of wholeness: the Universe, as well as some of its subsystems, form a totality whose overall behavior is richer than can be obtained from the sum of its parts.

---

[75] To measure the mass of a body, divide the force applied to it by the acceleration the forces produces. Equivalently, mass is the ratio of a body's weight to the gravitational acceleration g.

What is meant by Western physics? This is an important question because the answer determines whether physics can be taught, learned, or practiced in an Indigenous context, such as a classroom of Indian students, without conflicting with their identity or furthering cultural assimilation. Can Native students learn physics without betraying their tribal identity? I have also noticed the tendency to identify people with the culture that developed the academic discipline in which they work.

To answer this question, consider first the philosophy of mechanism, which is both a philosophy and an attitude. Physical theories in and of themselves are not mechanistic; nor are they atheistic just because they are not based on, or make reference to, religious faith. The one who formulates or applies a physical theory may or may not be a person of faith, but the theory itself is not religious or philosophical.

If by "physics" we mean the laws and theories that constitute physics, then it is meaningless to say that physics is Western or philosophically mechanistic. But it is meaningful to say that physicists (or other scientists) individually have mechanistic *attitudes* or *beliefs* based on the way they practice and interpret results. Viewing physics from a tribal perspective does not change the laws or theories of physics. To say that physics is a Western science because it was developed historically within Western culture is a mistaken idea. *Physics is not inherently Western.*

Regarding spiritual insight, physicists make up their own minds on how they interpret phenomena; they do not necessarily perceive the same realities the same way. When Bohm formulated an alternative approach to quantum theory that compares well to Indigenous views, he probably did not realize that he was close to the Indian circle of understanding. He was unconventional yet he never deviated from authentic physics. And we must not assume that spiritual insight is necessarily the result of intellectual brilliance. For a given researcher, a spiritual experience may occur anywhere and anytime, inside or outside the office or laboratory, or not at all. The sense of the sacred is not always equivalent to inspiration.

The broad premises of mechanism can be stated as follows:

- Everything consists of inert and changeless matter, which occupies space.

- The behavior of matter is inherently mathematical.

- All events in the Universe can be explained on the basis of Newtonian laws.

- The observer's mind is separate from the thing observed (the mind also being reducible to physical laws) and the ontological bridge between them are laws and theories.

It is safe to say that none of these premises agrees with tribal worldviews. Aristotle used reason alone to describe motion. Using reason alone and unaware of the equivalence principle (described in an earlier chapter), he proposed that heavier objects fall faster than lighter ones. In contrast, Galileo[76] proved Aristotle wrong by performing experiments and computing the acceleration of objects. He was the first to use mathematics to describe motion and is credited for introducing modern science. He also distinguished between velocity and acceleration.

Descartes, whose name is virtually a synonym for the Western paradigm of Cartesian dualism—the notion that two distinct kinds of things exist: mind and matter, or mind and body—suggested that the human body works like a machine. One attribute of matter is that it "extends"; it has boundaries and occupies space, whereas the mind (or soul) does not because it is immaterial. The mind and the body are two

---

[76] Galileo Galilei's last name is rarely referenced.

completely different types of substance; although they interact with each other, the mind cannot be identified with the body. Descartes reasoned that the body could be divided up by removing a leg or an arm but the mind cannot. The relationship between mind and body, particularly between consciousness and the brain, is known as the mind-body problem.

Mechanism can be extreme, extending to the idea that the whole of life, including what we call biology, can ultimately be understood in mechanical terms: DNA, genes, natural selection, and so on.

Chapter
15

THE MODERN WORLD

L iving in ways that do not require our attention to the hidden and remote forces that affect us is a choice. Even religion can be passive if there is no conscious interaction with the unseen and un-manifested world in our daily routine. Doctrinal beliefs and moral ideals may be strong but of little consequence if they are confined and protected within a small sphere of experience.

Each day brings new opportunities and challenges. There are always reasons for which to express my gratitude to the Creator and to the things he has made. There is always something for which to be thankful; beginning the day with a prayer helps my attitude later as I confront situations that will almost inevitably occur.

Being open to mystical and spiritual phenomena is commensurate with the notion that the Universe is alive. Modern physics is itself mystical; the many-worlds interpretation of quantum mechanics, the Schrödinger cat paradox, the phenomenon of non-locality, and wave-particle duality are all mystical ideas. Of these, only non-locality is a confirmed phenomenon but it is still strange.

Quantum and relativity theory have required fundamental changes in our understanding of space, time, matter, and other physical quantities. Much of the terminology that is valid in the classical realm is also used with reference to quantum events but the terms do not have the same meaning. Classical assumptions do not hold in the quantum realm. These can be summarized as follows:

- *Continuity.* Motion in the classical world is *continuous*: an object or particle moves through space along a continuous and well-defined path. At the quantum scale, motion is discontinuous due to the indivisible and instantaneous transitions that occur when an electron, for example, jumps from one orbit to another without going through intermediate steps. This indivisibility of motion also implies wholeness. General relativity

- *Determinism.* In the classical world, if we know with accuracy the position, velocity, and forces acting on every object in the system at a given point in time, the system's configuration at a future instant of time can be determined. (Also, because time can be reversed in Newton's equations, a system's past configuration can also be determined.) Quantum events cannot be predicted precisely; the rigid determinism in the classical realm must give way to approximate, random, and probabilistic event predictions.

- *Immutability* Whereas in the classical world elementary particles have a rigid "intrinsic" nature that does not change (it is *immutable*), in the quantum world they can behave like waves or like particles.

- *Reductionism.* The concept that the world consists of separately analyzable parts, each with a nature and existence that do not change, has been deposed. A common way of expressing this reductionist view is to say that "the whole is equal to the sum of its parts." In contrast, the non-mechanistic character of modern physics suggests that we think holistically: the world acts more like a single unit; the nature of every part depends on how it relates to its surroundings.

- *Locality.* The belief that an object is influenced directly only by its immediate surroundings must yield to the reality of non-locality.

- *Positivism, realism.* The positivist view that physics is only about things that can be measured no longer holds. Local realism—that all objects must have a pre-existing value before they are observed—does not hold (consider the EPR paradox, for example).

The classical world is not concerned with the philosophical questions about the nature of reality that arise in the paradoxical quantum world where direct observation is not possible. Even so, we must always bear in mind that the entire universe is sustained by an invisible world of random, *acausal* events.

What is actually present or what events occur in space and time that are independent of the human mind? Does it matter whether or not particles are considered to exist before they are observed? Particles are a part of nature, so the question is really whether a part of nature exists independent of human perception. The answer depends on our worldview because mind and perception are also part of nature. Nature has its own meaning to Native peoples.

The question of existence is an ontological one; it is about reality itself (what is actually there?), not what measurements can tell us, which is epistemological knowledge. We know from the objective evidence obtained through measurement that particles exist without assuming their existence in advance.

The question is important to physicists like Bohm who believe that quantum theory is concerned with what actually exists in the quantum realm, not only with what can be learned from measurements. From a Native perspective, I do not believe the physics question is important; whatever lies behind the things that are seen is revealed, if at all, through experience.

The quantum world has its own rules that defy Newton's laws. Physicists do not even speak of objects in the quantum world, only actions, events, and relationships. The cosmos is a dynamic web of relationships. Space and time are no longer conceived as existing independently; they have been replaced by space-time. Matter (mass) and energy are equivalent. A large mass changes the space-time geometry; in turn, matter must move according to that geometry. The length of an object is not constant; it changes when moving relative to an observer. Mass increases with increasing speed but cannot become infinite. And finally, if widely-separated particles can influence each other instantaneously, we have to wonder what else is factual and possible.

Hidden Realities

An active and participatory relationship with the natural world undergirds the practice of a science that is sustainable and should undergird everything we do. The counterintuitive nature of quantum theory, and in some ways relativity theory, calls for an escape from the common adherence to classical Newtonian notions that have been disproved. For thousands of years, the dictum "only what can be measured is real" has limited the quest for knowledge; by not recognizing the hidden realities revealed by modern physics, we further limit our knowledge and experience.

Those hidden realities involve the very small, the very fast, and the very remote. Gravity, energy, space, time, and matter are obvious but not well-understood realities. Space and time are not absolute and independent; they are relative, co-dependent, and change together in the form of space-time. Mass is an inertial property of matter; inertial and gravitational masses are equivalent. Mass and energy are also equivalent. And as I have already described, several properties of a physical body act as sources of gravity and are accounted for in Einstein's equations of general relativity.

Indian ways may also seem "unconventional" from a mainstream society's perspective but it is Western culture that has deviated from the natural order. Undoubtedly, many who agree in principle that living harmoniously with nature is the proper way to live will also claim that the development of Western civilization as it actually happened—except for imprudent mistakes that were made along the way—was inevitable on the basis of practicality; if civilization had not progressed, they would say, we would still be living in the "dark ages." But I do not believe that the original teachings given to the West were meant to be disregarded.

All things that are manifested ultimately spring from an invisible order. The concept of a primordial, active, transmutable substance is appealing to me because of its compatibility with the nature of space-time, the creation of matter during the Big Bang, if it happened, the imbuement of all things with spirit, and the manifestation of all things from a prior spiritual existence. The constant flux in the Universe is due to activity that occurs mostly at subatomic levels and involves the transformation of particles into pure energy and reverse transformations into matter but generally not into the same particles as before. These transformations involve transient virtual particles that exist for incredibly small moments.

The following categories of events represent situations that cannot be directly observed with the unaided eye and do not consciously affect our daily lives:

1. The motion of quantum particles

2. Transformations of quantum particles

3. The motions of planets, galaxies, and other remote celestial bodies

4. The motion of objects traveling close to the speed of light

We are affected in different ways by what occurs in each of these realms but we are not compelled to act in a conscious manner in order to maintain a practical existence. As an aid to understanding the differences in physical theories based on differences in scale, physics can be divided into three domains connected with the theory that applies to each:

Category	Domain	Theory
A: large	Planets, galaxies, etc.	General relativity
B: small	Quantum scale	Quantum, M-theory
C: every day	The visible world	Newton's laws

General relativity actually applies to all scales except the quantum scale. *Quantum cosmology* applies quantum theory to the entire Universe. *Quantum gravity* refers to various approaches that attempt to unite Einstein's theory of gravity (general relativity theory) with quantum theory into a single theory.

Referring to categories A, B, and C and cases 1 through 4 above, the following table shows how they are related.

Category	Theory	Case
Physics of the large	General relativity	3, 4
Physics of the small	Quantum, M-theory	1, 2, 4
Physics of every day	Newton's laws	

Quantum theory is so strange that, for people who are first exposed to it, it is like "nothing they have ever seen before."[77] Neils Bohr, the physicist who first proposed a model of the atom consisting of a nucleus at the center with electrons moving in circular orbits around it, said that quantum theory is "shocking" when first encountered. Common conceptions of time, space, and gravity are erroneous, according to the discoveries of relativity, which have been validated experimentally. Likewise, some concepts of quantum theory are counterintuitive and incredible but have been experimentally confirmed. While these statements are made with confidence, I hasten to point out, as eminent physicists including Einstein and Bohm have explicitly stated, that physics theories are tentative and valid only within their domains.

The list below summarizes the main concepts of modern physics.

1. Time and space are mutually dependent, not absolute and independent, entities. Events do not occur in space and time but in space-time. Space and time both change in the presence of mass. Space-time geometry resembles a flexible fabric or web.

2. The Universe is an unbroken, coherent whole whose complexity has not been fully revealed. Since it is not reducible to particles of matter, the ultimate constituent (so far) appears to be the universal energy that encompasses the Universe.

3. Newtonian physics applies to the visible domain of common experience. Bohm calls this domain the *explicate order*.

---

[77] Richard Feynman, a renowned physicist who among other things is known for helping develop the atomic bomb, is known to have made this statement.

4. Quantum physics applies to the invisible world of atoms and subatomic particles and processes. Bohm calls this unmanifested world the *implicate order*.

5. Quantum reality lies with the relationships within the flux of energy and processes of quantum nature (Heisenberg). The co-existent particles and energy undergo ceaseless transformations.

6. The Universe consists of processes, not static things influenced by causal forces. Modern languages, including English, are inadequate to describe quantum processes. Subjects acting on objects through a verb cannot describe this world. On the other hand, Native languages are action-oriented and suitable for describing the quantum world, a fact that Bohm discovered shortly before his death. Processes can best be described by telling a story.[131] Nature at all scales is in a state of constant flux.

7. Matter and energy are equivalent; energy always involves movement, including the internal energy due to the motion of particles within every object.

These concepts do not require us to invoke metaphysics; in particular, the physical wholeness of the cosmos in itself does not have a spiritual implication but neither does it have a secular one.

Cosmic Wholeness

The classical world of daily experience consists of visible objects and particles of all sizes. Whereas in classical thought, the Universe functions like a machine, the non-mechanistic character of modern physics tells us that the world is an undivided whole.

Physicists working on M-theory and quantum gravity describe the Universe as a "network of relationships" and a "seamless holographic fabric"—terms that imply wholeness. A network consists of nodes and links (threads, relationships, or electromagnetic links such as in the worldwide web) in which any two nodes can communicate with each other through one or more "jumps." A network is a single, unbroken entity. Bohm and other physicists have been convinced that the Universe itself is a giant hologram.

The Universe is in constant motion and transformation at all scales. Atoms change quantum states when electrons jump between orbits without passing through intermediate states, implying that motion is indivisible.

Wholeness is further manifested, of course, in the phenomenon of non-locality, an undeniable feature of wholeness and connectedness in the Universe. Both quantum theory and relativity theory give strong evidence of wholeness in the cosmos. Quantum theory suggests wholeness in at least the following ways:

- in the indivisibility of motion at the quantum level (an electron can go from one state to another without going through intermediate levels);

- in the non-local connectedness of entangled particles;

- in that the physicist becomes more intimately related to what is being measured. The observer and the observed form a single system, although they reside at different levels of reality—a relationship which does not exist in the classical order of reality.

With regard to general relativity, we first note that Einstein envisioned a Universe with no rigid bodies—three-dimensional objects with definable boundaries; instead, these objects consist of locally intense concentrations of an encompassing gravitational field. The gravitational field acts like a web in which particles are abstractions. The theory further reinforces the concept of unity and wholeness in the cosmos because there are no discontinuities. In a practical sense, things that are visible ultimately consist of invisible particles that are being transformed into pure energy and vice versa. Objects maintain their shape because of a renewal process that is constantly occurring at the sub-quantum level. General relativity theory describes an order of existence that differs radically from the classical world. As already mentioned, quantum theory further strengthens Bohm's view of "undivided wholeness."

Bohm developed a cosmology from quantum theory based on undivided wholeness using the same premise for his theory that Indigenous peoples have always incorporated into their ancient traditions, not as dogma but as an applied concept. Bohm's picture of an enfolding/unfolding universe is so similar to the cosmology of the people of the Andes, as reported by Oakley Gordon from the University of Utah, that it's worth mentioning here.[132] In his paper, "An Environmental Epistemology of the Andean People of Peru," he tells how he arrived at an approximate understanding of their cosmology after listening to a spiritual leader and teacher, Américo Yábar, over a period of several years. In metaphoric terms, the cosmos is seen as a vast "web" of filaments of energy, which could also describe a gravitational field if massive objects are close to each other. Points where they come together form a node, and each node represents what is experienced as an object. In this model, the Universe is a single, unified entity, everything is interconnected through filaments, and there is little or no distinction between animate and inanimate (these are non-Indigenous terms). The purpose of metaphor, Gordon explains, is to aid in understanding an epistemology other than one's own, especially if the other one is based on experience. In reality, one cannot adequately describe another epistemology in one's own language, because language co-evolves with the epistemology; the language belongs to it.

Bohm also points out that modern languages, including English, cause the totality of existence to become fragmented with their subject-verb-object structures. Human thought is forced to accommodate the subject as a separate entity, which is also true of the object if the verb is transitive. In contrast, if the world is viewed as a coherent whole, the language should reflect the unbroken two-way flowing movement between thought and experience. Quantum processes are action-based; they require such a language. To correct this shortcoming, Bohm created a new mode of language, which he called the *rheomode*, based on English, that gives the verb a primary role. However, as physicist David Peat tells us (*From Certainty to Uncertainty*), Bohm later met Natives who were fluent Algonquin-language speakers shortly before his death and discovered that their language had the feature he was looking for—one that connected language and worldview—to describe quantum physics.[133]

Werner Heisenberg, known for the Uncertainty Principle, argued that quantum reality cannot be represented by electrons and protons acting as "building blocks" of matter. They are visible manifestations of quantum processes based on relationships that exist within the flux of energy and processes of quantum nature. Particles are always in a state of flux (between matter and energy) and do not always transform themselves back into the same particles.

To illustrate the difference between holistic and fragmented views of the world, Bohm used the example of a functioning watch consisting of components connected together. If the watch is carefully taken apart, it can be reconstructed by noticing how each component relates to all of the other components. However, if the watch is shattered with a hammer, the result is a pile of *fragments* that no longer reveal the relationships among components, making it impossible to reconstruct the watch. The conventional (fragmented) approach of Western science is similar to the shattered watch; systems and subsystems in nature are studied as if they did not relate to each other.

Einstein's equations are based on field theory instead of particle theory. For a long time, die-hard physicists held on to an atomistic view of the Universe, even hoping to explain light using Newton's laws. By continuing to adhere to Newton's laws, they were also holding on to a mechanistic view. However, certain light phenomena, such as diffraction and interference, can only be explained with wave theory; a particle theory of light cannot explain it. Field theory also accommodates wave motion in which space-time is the "glue" in the background relative to which all motion occurs including gravity traveling through it at light speed.

The physical wholeness in the cosmos thus corresponds to the well-known phrases, "we are all related" and "all things are connected," which are often expressed by American Indians with reference to all humans and non-humans. To Indigenous people and all who have come to understand it, this wholeness speaks to the sacred web of life. The Indian world prior to European contact was complete under the preeminence of natural law.

The Quantum World

Amir Aczel, introduced earlier, said that quantum mechanics is the strangest field in science and non-locality the strangest phenomenon in physics. He described non-locality this way: "Two particles that may be far apart, even millions or billions of miles, are mysteriously linked together. Whatever happens to one of them immediately causes a change in the other one."[134] The two particles are said to be "entangled."

Einstein and others had questioned non-locality—a predictive feature of quantum mechanics—since 1935 but is no longer just a possibility, a probability, or a postulate. It is a confirmed phenomenon that seemingly out-performs light and gravity because it acts instantaneously. Numerous experiments conducted in the 1990s and later have vindicated the predictions of quantum theory. Besides being bizarre, one might think that non-locality implies faster-than-light communication, which is not possible. However, the twin entangled particles correlate their behavior without communicating through signals as if belonging to a single system, regardless of the distance between them. Non-locality is a feature of the physical world, the full meaning of which is something to ponder; it gives evidence of an undivided Universe.

Physicists cannot be expected to agree on the interpretation of natural phenomena either among themselves and especially compared to an Indian trained in science. Nor can we assume that spiritual insight stems from intellectual brilliance. However, important discoveries in quantum theory did involve a certain amount of instinct and inspiration, as well as intellectual ability, without relying strictly on the empirical methods of science. The phenomena discovered by Heisenberg, Schrödinger, Planck, de Broglie, Bohr and others are relevant to Indian thought because they invoke several questions about the validity of experience, consciousness, reality, time and space, the enfolding and unfolding nature of energy and matter and its relationship to the seen and unseen worlds, the relationship of energy to spirit, and so on. Philosophical questions about the nature of reality are unavoidable in quantum theory.

The transformations which occur at the quantum level consist of subatomic particles too small to be visible to the naked eye and even with a microscope. Physicists can detect invisible particles and study their behavior by examining recorded traces of scattering events (particle collisions or near collisions) produced by particle accelerators. Any attempt to measure a quantum system will disturb it because the wavelength of the light used for probing approximates the wavelength of the particles being measured.[78] The accuracy in measurement is limited by the Heisenberg *Uncertainty Principle*; position and momentum cannot both be precisely determined simultaneously.[79]

How do we know the particles are real (*i.e.*, they actually exist) if all we see are their traces and how do we know that the traces correspond to those particles? A positivist believes that only authentic knowledge obtained empirically is real. A *logical positivist* believes that both observational evidence is necessary *and* it must be verified. However, non-locality contradicts this rule because it is a phenomenon that was confirmed through inference, not direct observation; what *is* observed are the correlations. In any case, the positivist position has to be abandoned.

When Heisenberg first tried to explain the Uncertainty Principle in classical terms, Bohr retorted that the electron does not have intrinsic properties such as position and momentum. This is not only a philosophical issue but also one of ambiguity because, although classical terms like *wave, particle, momentum, position,* and *trajectory* have precise meanings in classical Newtonian theory, it is not true in quantum mechanics.

For example, a particle can behave like a particle and like a wave; a particle has position but a wave does not. Heisenberg still thought of a particle as possessing well-defined properties

---

[78] Here is a helpful analogy: Think of a boat floating on water. Small ripples do not disturb the boat, but if the boat were reduced to the size of the ripples, it is easy to see that the boat would bob up and down.

[79] The mathematical expression for this principle is $\Delta p \Delta x \geq h/4\pi$, where p is the momentum, x is the position, and h is the Planck constant.

like position, momentum, and trajectory. This view provoked Bohr, who adamantly insisted that the precise path of a particle should not be called "uncertain," implying that the particle has definite properties before they can be determined. Rather, he said, it should be called *ambiguous*, just as temperature is inherently ambiguous because it is a measure of the amount of kinetic energy of an ensemble of molecules but temperature has no meaning for individual molecules. The problem, he said, lies in the fact that the entire phenomenon (quantum system plus apparatus) cannot be analyzed at the lower (quantum) level of detail. In the same manner, he argued, it makes no sense to talk about position, path, or momentum as if they were real entities.

In his old age Heisenberg expressed the view that any talk about electrons and protons as "building blocks" of matter was a confused misrepresentation of the nature of quantum reality. Rather, he said, they are the *surface manifestations* of underlying quantum processes. Symmetries are the more fundamental property instead of particles. Reality lies not with the particles but with the *relationships* that exist within the flux of energy and processes of quantum nature.

Particles are always in a state of *flux*, and this is yet another characteristic of the quantum world. A particle can be transformed into energy and energy can transform back into a different particle. *Nothing in the Universe is at rest;* this would violate the principle of uncertainty, because momentum (zero) and position would be precisely measured. The quantum world is dynamic, which implies that the entire Universe is undergoing constant change. Everything consists of atoms, subatomic particles, and ultimately only processes. The Universe itself is constantly expanding and the rate of expansion is growing.

Returning to wave-particle duality and the question whether a particle can act as a wave simultaneously, the two

The Condor and the Eagle

forms can only exist separately, according to the conventional interpretation of quantum theory. This leads to the paradox associated with the double-slit experiment in which a single photon (particle) will produce wave patterns on a screen after it passes through the slits. But how can a *single particle* pass through *both* slits, as the conventional quantum theory claims? Although it doesn't make sense, the conventional interpretation accepts this bizarre characteristic of the quantum world. The belief that particle and wave cannot co-exist is known as *complementarity*, another one of Bohr's philosophical notions.

Bohm, however, offered an alternative explanation, saying that the particle always behaves, in certain cases, like both particle and wave together. When light approaches a double-slit, it does not decide to become wave or particle but passes through as both slits. Even in the case of a single photon, the waveform passes through both slits while the particle goes through only one of the slits.

In Bohm's view, the conventional Copenhagen interpretation limits physics. Conventional quantum theory does have an accurate method of computing the probabilities of experimental results, but it does not explain quantum processes. "Indeed, without the measuring instruments in which the predicted results occur, the equations of the quantum theory would be just pure mathematics that would have no physical meaning."[135] In Bohm's formulation of quantum mechanics, a quantity of major significance, which is missing from the conventional one, is the *quantum potential*. This appears as an extra term in the Hamilton-Jacobi equation of quantum mechanics and acts in addition to the potential[80] in the original equation. Its value does not diminish with distance, a property that corresponds to the observed phenomenon of non-locality; it

---

[80] Potential is stored energy due to position.

influences everything and it is also affected by the surrounding field. It allows a particle, such as an electron, to respond actively, not passively, to the potential's presence. The quantum potential seems to allow a particle to have its own "awareness." Non-locality is explicit in Bohm's formulation. Considered together, these properties imply wholeness and interconnectedness in the Universe.

Einstein disagreed with Bohr not only on the issue of ambiguity but also on the meaning of reality. The issue led to a serious dispute between them, yet many believe that the central issue was the use of language. Once close friends, the two men eventually separated. The rift was so deep that, when a mutual colleague arranged a party so that they would associate and come together, Bohr congregated with his students and Einstein with his.

Max Planck's discovery of the quantum involved inspiration, instinct, or whatever a stroke of luck implies, but it was not deduced by the formal scientific method described in textbooks. The story, briefly, is as follows: Until 1900, the light from heated bodies could not be explained using the mechanical and continuous theory of light.[81] Newtonian theory wrongly predicted that a hot poker would give off its electromagnetic energy[82] continuously at frequencies beyond the ultraviolet. Planck proposed a purely theoretical hypothesis, saying that the energy is emitted *discontinuously* in lumps of energy, or *quanta*. This "crackpot" idea had no basis in a mechanical Universe; it was completely unexpected, a "lucky" guess. But it correctly related the energy E (given to the wave by the oscillating material) to the frequency f of that wave ($E = hf$, where h is the

---

[81] Classical mechanics (Newtonian physics) assumes that motion is continuous, for it was inconceivable that an object could move, or jump, instantaneously between two points without moving through all of the space in between.

[82] Light is a form of electromagnetic energy, including radiation from a hot object.

Planck constant). This meant that light waves do not behave like mechanical waves. The very small constant h, and therefore the quantum world, is incredibly small. Also, each quantum consists of whole photons; fractional photons, such as ½ hf, cannot exist.

The quantum world is nothing like our world of common experience, yet everything we see is manifested from an invisible order. In the classical order, motion and equilibrium are the result of forces applied to objects, and motion is continuous. But in the quantum world, motion is discontinuous and random, not the result of known forces. When an electron in an atom jumps *instantaneously* between orbits, it releases or absorbs energy in packets; these quanta consist of photons, each possessing energy in packets, an integral multiple of hf. When an electron jumps to another orbital level, the atom changes its energy state. Thus, in the quantum world, motion is unpredictable, discontinuous, non-causal (random), and uncontrollable—a behavior that is the opposite of classical behavior. Furthermore, each atom in nature has a unique set of energy states, giving it a unique footprint.

Unsurprisingly, quantum theory also radically changed the relationship between observer and observed, because the observer (physicist) is now in close contact, so to speak, with the quantum system that is being measured, thus increasing the subjectivity of the experience, albeit the quantum world is not directly observable.

It seems that the farther we remove ourselves from what we observe, the less social toward it we become, which may explain in part the predominantly unsocial attitude toward nature when we work in the physical sciences compared to the life sciences. Likewise, a mechanistic philosophy may diminish our empathy for all forms of life. These are questions to ponder.

Quantum Processes

Imagine a world where there are actions but no objects, only processes and relationships. There are no forces acting on the objects and events occur randomly, so there is no cause and effect. A human language with a subject-verb-object structure, such as English or another modern language, is inadequate for describing this kind of world. Native languages, as already mentioned, are action oriented and suitable for describing quantum processes.

Native languages can describe a tribe's dynamic environment. It is ironic that the world's dominant science developed over millennia using languages that cannot describe the fundamental processes in the cosmos. Native languages not only have this ability but they also reflect knowledge of the Universe which is itself dynamic. Native languages are born from the land. "The land taught us our language," says Jeanette Armstrong, a fluent speaker of her Okanagan language in Washington state. I've seen tribal friends and acquaintances struggle to express certain thoughts in English.

The inability to directly observe quantum events when they occur has led to the unresolved problem of wave function collapse known as the "measurement problem." Particle collisions that occur at high velocities can be examined only after the event occurs, giving rise to variant interpretations among physicists. Unlike Newton's equations that describe the classical realm, the equations of quantum mechanics do not predict what *will* happen, only what *may* happen out of a range of probabilities given by the wave function.

Classical measurements are objective because the observer and the object under observation exist at the macro level. In quantum physics the status of the observer (the physicist) has been redefined. The influence that the process of observation has on the system observed is significant. In quantum physics

The Condor and the Eagle

the physicist does not merely observe; he/she has to participate interactively in the measurement process and thus influences the observations of quantum events. The subjective and integrated nature of the measuring process exemplifies holism, arising from the need to take the whole physical system—the measurement apparatus acting as the observer and the object being measured—into account. The physicist as observer (through the measuring apparatus) becomes a part of what is being measured; they form a single, unified system. The physicist is no mere passive witness, as in the classical case, but an active participant. The new relationship implied by the connectedness between observer and observed has been referred to as "quantum wholeness."

Quantum wholeness is only one form of wholeness revealed by modern physics. Relativity is yet another form because it is a field theory in which space, or rather space-time, acts like a web that encompasses all space. Another interesting fact is that, in relativity, the most basic measurements, such as the mass, velocity, and length of an object, are no longer constants as in the classical world but depend on the frame of reference of the observer. Every observer sees something different.

Another example of wholeness is a bound electron within an atom, which has to be considered part of the state of the whole atom and the electron loses its individuality. Because of *Pauli's Exclusion Principle*,[83] when another electron is added to an atom, it has to assume a different state from that of any of the electrons already present. No two electrons can be in the same state in the same atom. One practical outcome of the Exclusion Principle is that matter as we know would collapse without it.

---

[83] Wolfgang Pauli was one of the eminent pioneers of quantum physics. Electrons, protons, and neutrons are all subject to this principle, and atoms occupy space in accordance with it.

Every physicist or other scientist has the opportunity to come face-to-face with nature, so to speak, either conceptually through the interpretation of data or more directly through a personal contemplation of the cosmos. In the domain of distant and massive objects in the cosmos, the Hubble telescope has provided incredible images of stars, galaxies, and nebulae. How we interpret what we see far away in the sky or perceive close to us depends on our individual perspective.

Chapter
16

SIGNS OF CHANGE

According to the latest physics research, the Universe is an interconnected network, like a tapestry or web. This perception of the cosmos is relatively new to the world of physics but a very old principle embraced by Native peoples. From the viewpoint of contemporary science, physics is also the study of "inanimate" nature, somewhat like the atomists of the 5th century BC who theorized that the natural world consists of two fundamental components—*atoms* (indivisible bits of matter) and the *Void* (a vacuum, empty space)—and that atoms are supposedly indestructible and immutable.

Although today we know that atoms are not indestructible or indivisible, nature at its most basic level is still generally perceived by Eurocentric scientists as inert and passive. However, a mechanistic theory of nature—the premise that the Universe functions like a watch or a machine that can be reduced to the smallest constituent particles of matter, or "building blocks," and then reassembled—is no longer considered a sensible view. Native peoples, of course, hold that nature is dynamic and alive.

Sometimes a new idea is conceived or discovered to possess a deeper meaning within an existing idea, but a closer examination reveals that it is not actually new. This is the case with "deep ecology," mentioned in chapter 6, which is purported to be a new way of thinking about our relationship to Earth but is actually a very old traditional belief and practice of tribal peoples. A similar case is physicist Nick Herbert's notion of quantum animism that "presents the idea that every natural system has an inner life, a conscious center, from which it directs and observes its action."[136] I haven't examined the concept in depth but it is a familiar one. A positive way of looking at these "new" insights is that good-thinking people are confirming what Natives have always known..

Holistic Approaches

There are encouraging signs that the world seems ready to listen. Attitudes and approaches involving scientific research conducted across various academic disciplines are changing in recognition of the holism in nature. Since the advent of environmental science, which focuses on human-caused pollution and degradation of the world we inhabit, scientists have begun to accept the more realistic view that Earth is a single interacting system.

Some academic disciplines are being combined into a larger domains, providing a compelling example of a shift away from reductionism. Examples include complexity theory, chaos theory, general systems theory, interdisciplinary doctoral degrees, environmental studies, and climate change research. These interdisciplinary areas recognize the interrelated and interconnected nature of systems and networks which function without human intervention. Western science will likely continue to validate and hopefully implement some of the ancient knowledge that our ancestors practiced since long before contact with Europeans.

General systems theory (GST) was originally proposed in 1928 by biologist Ludwig von Bertalanffy, who also coined the term. Until reductionism was finally discredited, the scientific method, which embraces Cartesian dualism, was rarely questioned. In fact, the required textbook for a physics course I taught in 2000 described this method in detail and highly praised it. Bertalanffy repudiated the notion that a physical system can be reduced into individual components, each existing and functioning as an independent entity, and then used to describe the whole system. On the contrary, he argued, the components of a system inherently interact with each other and, furthermore, the interactions are nonlinear. [84] GST was extended a few years later to include biological systems. An important implication of GST is that it recognizes the wholeness in nature, including the wholeness in living systems, and is successful in making predictions because of it.

Bertalanffy not only recognized the inherent holism in natural systems but also wanted to advocate an ethical view, considering the dehumanizing effects of prominent mechanistic thinking. His mathematical model of an organism's growth over time, published in 1934 in collaboration with other researchers, where mechanistic thinking is particularly critiqued, is still in use today. Since the end of the Cold War, efforts to strengthen an ethical view have helped to bolster an interest in general systems theory.

GST recognizes holism in nature. Bertalanffy stated that systems cannot be understood by investigating their respective parts in isolation. On the nature and aim of GST, he said:

> Not only are general aspects and viewpoints alike in different sciences; frequently we find formally identical or isomorphic laws in different fields. In many cases,

---

[84] Nonlinearity was described in chapter 3.

isomorphic laws hold for certain classes or subclasses of 'systems', irrespective of the nature of the entities involved. There appear to exist general system laws which apply to any system of a certain type, irrespective if the particular properties of the system and of the elements involved. General systems theory is, therefore, a general science of wholeness.[137]

GST recognizes not only the wholeness in nature, as discoveries in modern physics have revealed, but also the wholeness in living systems—a principle reflected in the equations of GST.

Chaos theory was born in the 1960s when a mathematical weather model yielded surprising results. Meteorologist Edward Lorenz noticed that his model predicted a *very large* change in the weather over time from a *very small* difference in the initial conditions. This kind of behavior is a fundamental characteristic of chaos theory studies. Chaos theory encompasses several disciplines—engineering, economics, biology, philosophy—and is used widely in the natural sciences and more recently in the social sciences. In technical language, chaos theory studies the unstable aperiodic behavior of deterministic nonlinear systems.

A stable system, like a rocking chair, tends to move toward a point of equilibrium; when a force is applied in a direction away from equilibrium, the system reacts with a force that tends to bring it back into its equilibrium position. As the chair rocks, it gradually vibrates in shorter forward-backward movements past its equilibrium point until it comes to rest at that point. But unstable systems like those found in nature are often unpredictable; *i.e.,* non-deterministic.

Animals, physical systems, quantum systems, and nonlinear equations can exhibit unpredictable behavior. The interesting cases of unpredictability in nature are those which are sensitive to initial conditions, as in the above case of weather. Another example that introduces students to chaos theory is the double pendulum, which consists of two rods coupled end-to-end; the free end of one rod is hung from a stationary support as a pivot point; the other pivot point is the coupled connection. To start the motion, the coupled point is raised to a certain level (this is the initial starting point) and released so that the double-pivot system will swing; the starting point is called the "initial condition." If this is repeated several times and the movement of the free end of lower rod is recorded, say on paper or film, there will be a very *large* difference between records from *small* changes in the initial position. Also, whereas the motion of a pendulum consisting of a single rod is predictable, the motion of the double pendulum system is not, in view of the coupling's random motion. Randomness and unpredictability go together.

Somewhat linked to chaos in terms of behavior is *complexity theory*. In a complex system, many independent elements continuously interact and spontaneously organize and reorganize themselves into more and more elaborate structures over time. In natural systems with *emergent properties*, the component parts of the system are able to function together and self-organize in ways that they would not function individually. As with chaos, the behavior of self-organizing complex systems cannot be predicted and their components cannot be studied in isolation.

Paul A. LaViolette, with a background in general systems theory and physics, developed and published a physics methodology during the late 1970s based on the concept that *process* rather than *structure* is the basis of physical existence. He postulates that matter emerges from a primordial ether that fills all space in the universe. This is not the luminous ether that was hypothesized in the 19th century and rendered superfluous by

Einstein's paper on the special theory of relativity. Nor is this ether inert, passive, or unchanging; rather, it is postulated as an active transmuting substance at the base of existence that acts similarly to interacting chemicals in a reactive solution.[85] LaViolette has written several papers to demonstrate how, under this new paradigm, he is able to account for electric, magnetic, gravitational, and nuclear forces; it also resolves many long-standing problems in physics.

A major change in science as a whole will depend on the larger scientific community to make the right choices. However, modern physics, already called the "new" physics because its concepts transcend conventional thought, provides the most significant evidence of holism and, in my view, will be the greatest agent of change to a new paradigm in science that seamlessly accommodates the quantum and classical worlds, avoiding or preventing dichotomies on its way to acceptance. How it will relate to a theory of quantum gravity—that long-sought theory which will reconcile quantum theory and general relativity theory—remains to be seen.

It has been my contention that Western civilization contradicts the very holism and authority which is inherent in nature and is revealed by modern physics. This, of course, assumes that there is a relationship between a correct understanding of nature and the proper way to live and, moreover, that humans are obligated to relate to nature in ways that are compatible with that understanding. The beauty of human cultures is that each one is able adapt as necessary, since we are all connected through and dependent upon the earth, for even the ancients knew the basic elements necessary for a

---

[85] Reaction-diffusion is a process in which two or more chemicals diffuse over a surface and react with one another to produce stable patterns. Reaction-diffusion can produce a variety of spot and stripe patterns, much like those found on many animals. Developmental biologists think that some of the patterns found in nature may be the result of reaction-diffusion processes.

The Condor and the Eagle

continued physical existence: earth, air, water, and solar energy (which they called "fire"). The holistic practice of physics would re-align science into a proper perspective relative to the earth and the cosmos; the quantum's strange behavior has already been demanding explanations and interpretations that are beyond conventional physics. By recognizing the limitations that the scientific method imposes, a new or revised paradigm would extend the current domain of knowledge without changing physics or imposing religious or philosophical views.

A holistic view of physics has broad implications. Scientific knowledge as it is currently defined is incomplete knowledge. A re-alignment of science is necessary so that Western epistemology will reflect all of the endowments that we humans have been given—spiritual, intellectual, emotional, and physical—in the contribution to knowledge. In a way, it will repair the rift that occurred in the 5th century BC among Greek philosophers.

The Anthropic Principle

According to some astrophysicists, life in the Universe would have been impossible if some of the physical constants and other conditions in the early Universe had differed ever so slightly in the values they had, suggesting that the Universe is "fine-tuned" for the possibility of life. Among those who agree that the Universe is an unexpectedly hospitable place for living creatures to make their home, and seems to agree with this view, is physicist Freeman Dyson.

Theoretical physicists recognize the evidence of complexity in the cosmos. They do not deny, against all logical expectations, that the Universe is surprisingly friendly to complex structures and, amazingly, supports intelligent life on Earth as well as systems as large as galaxies. Unsuccessful efforts thus far to develop a complete theory that unifies all of the fundamental interactions (forces) of nature has caused frustration among cosmologists who have worked arduously for at least three decades. Some string theorists, convinced that the only remaining alternative is to accept the incredible complexity of nature as unexplainable, have come to accept the *Anthropic Principle* as inevitable. This principle can be expressed in several ways, such as "A human being, as he/she is, can exist only in the Universe as it is," or "Humans should take into account the constraints that human existence as observers imposes on the sort of Universe that could be observed."

One of the interesting things about being an "academic" in science is that you first assume that the physical world consists of random processes which also originated by chance. This is what society expects of us if we work in a public institution. Consequently, when we find evidence suggesting that there is purpose in nature or that the world came into being by some form of intent, we are expected to play the role of a skeptic and place the burden on others to argue otherwise but on *our* terms; *i.e.*, on the basis of empirical science. If they argue on *their* terms—metaphysics, mysticism, spirituality, etc.—we are not expected to consider their arguments valid because they argue on some other basis except empirical science. In this regard, the freedom I experienced in a tribal college environment was a great relief.

Anthropocentrism and Liberation Theologies

When we focus on human liberation by assuming that we have the right to enjoy all of the wealth we can accumulate, we tend to take more than we need yet never enough to be satisfied. This leads to oppression not only of people who are less privileged but our non-human relatives as well.

Total liberation obliges us to work in behalf of all life. "Environment," an anthropocentric term, implies that the natural world surrounds us, as if we were separate from that which sustains us, and that the world's goods exist to be exploited for our purpose. The environmental movement does acknowledge that humans are part of nature but it does not challenge the problem of greed and exploitation. We need a strategy that obliges us to act consistently with a sacred principle.

"A strategy for survival must include a liberation theology," say the authors of *A Basic Call to Consciousness*, who explain that liberation theologies are belief systems that challenge the assumption that the earth is a commodity to be exploited thoughtlessly; such a theology will also develop in people the consciousness of the sacredness of life, which is the key to human freedom and survival. "The renewable quality— the sacredness of every living thing, that which connects human beings to the place they inhabit—that quality is the single most liberating aspect of our environment."[138]

A New Physics Paradigm

In the 1970s, Paul. A. LaViolette introduced a new paradigm in physics which uses a general systems approach and is based on a concept recognized by ancient non-Western peoples, including the Egyptians. It postulates that matter emerges from a primordial ether that fills all space in the Universe. This new paradigm has been shown to account for electric, magnetic, gravitational, and nuclear forces, and it claims to resolve many long-standing problems in physics. I know of no other approach in physics which incorporates the principle that matter emerges from a live substance.

Generally speaking, an ether is a space-filling substance or field thought to be necessary for the propagation of electromagnetic waves. An ether was proposed in the time of Einstein as the medium necessary for light to propagate.

Einstein did not have to disprove the ether hypothesis; his paper on the special theory of relativity simply rendered it superfluous. Interestingly, toward the end of his life, he proposed that the space-time of general relativity theory be considered a type of ether.

LaViolette, who has written several books and published many original papers in physics, astronomy, climatology, general systems theory, and psychology, developed this paradigm using an ether concept that is genuinely different from the luminiferous ether that was hypothesized in the time of Einstein. His theory of *subquantum kinetics* (SQK) is a revolutionary physics methodology founded on the concept that *process*, not structure, is the basis of physical existence. This is a very important difference; the theory assumes that matter emerges, not from an inert, passive, or unchanging ether, but from an active and transmuting ether, similar to chemicals in a reactive solution. [86] "This approach postulates an active, interactive subquantum substrate whose processes give birth to and continually sustain the physical form that makes up our Universe," says LaViolette.[139]

For his revolutionary idea in foundational physics, LaViolette credits Nikola Tesla and Alfred North Whitehead's organic theory of space, which describes an ether that is active, interactive, and "alive." According to Whitehead, space is a kind of living organism in which the whole of space is *more than the sum of its parts* and every volume of space interacts with every other volume.

Tesla was an inventor and an important contributor to the birth of commercial electricity but is perhaps best known for his

---

[86] In a reaction-diffusion process two or more chemicals diffuse over a surface and react with one another to produce stable patterns. Reaction-diffusion can produce a variety of spot and stripe patterns, much like those found on many animals. Developmental biologists think that some of the patterns found in nature may be the result of reaction-diffusion processes.

revolutionary developments in the field of electromagnetism in the late 19th and early 20th centuries. He wrote a minimum of 278 patents. He said (c. 1930):

> Long ago he [humankind] recognized that all perceptible matter comes from a primary substance, or a tenuity beyond conception, filling all space, the Akasha or luminiferous ether, which is acted upon by the life-giving Prana or Creative Force, calling into existence, in never ending cycles, all things and phenomena. The primary substance, thrown into infinitesimal whirls of prodigious velocity, becomes gross matter; the force subsiding, the motion ceases and matter disappears, reverting to the primary substance.[140]

Innovative and revolutionary thinkers like Tesla are often shunned and discredited for their seemingly outlandish proposals. It is said that because of his eccentric personality and his seemingly unbelievable and sometimes bizarre claims about possible scientific and technological developments, Tesla was ultimately ostracized. Fortunately, video documentaries like *Nicola Tesla: the Genius Who Lit the World, The Secret of Nikola Tesla*, and others have recorded some of his accomplishments and are still available.

Although there are no Indigenous sources to my knowledge that explicitly convey the concept of a primordial ether, the concept seems compatible with the notion of pervasive energy, or creative Power, and with "all things are imbued with spirit," and the notion the Universe is alive.

LaViolette also quotes Egyptologist R. T. Clark on the notion of a primordial ether that is found in Egyptian creation myths:

> Every [Egyptian] creation myth assumes that before the beginning of things the Primordial Abyss of waters was everywhere, stretching endlessly in all directions. It was

not like a sea, for that has a surface, whereas the original waters extended above as well as below. They are the basic matter of the Universe and, in one way or another, all living things depend upon them.[141]

"Nature is dynamic and transitory in its subatomic essence and constantly renewing itself," says LaViolette. "The theories of physics developed from the top down, whereas SQK begins at the bottom from the underlying, invisible, and unobservable ether."

THE WEB OF LIFE

So do you not feel that, buried deep within each and every one of us, there is an instinctive, heartfelt awareness that provides - if we will allow it to - the most reliable guide as to whether or not our actions are really in the long-term interests of our planet and all the life it supports? This awareness, this wisdom of the heart, may be no more than a faint memory of a distant harmony rustling like a breeze through the leaves, yet sufficient to remind us that the earth is unique and that we have a duty to care for it.[142]

–Charles, Prince of Wales

Long before the World Wide Web, storytellers told about the creative powers of Spiderwoman and Grandmother Spider and the web of life. The web they weaved connected human beings to each other and to the whole cosmos.

In holistic terms, the world acts more like a single entity; the nature of every part depends on how it relates to its connections and surroundings. Imagine a world that takes these interdependencies seriously, where everyone and everything is given due respect according to purpose and the right to exist. The early Greeks emphasized purpose in their view of the world, until a new mode of thought was introduced and debates ensued.

Wholeness implies the interconnectedness, interdependence, and inter-relationship of all things. As modern physics has revealed, these qualities are inherent in the cosmos but physics does not oblige us to act accordingly. This obligation comes from a land ethic, which assumes that there is a relationship between understanding nature's laws and how we conduct our lives. A land ethic based on these understandings implies reciprocal relationships: where there is influence, respect is returned; where there is interdependence, there should be a caring concern for others; interrelatedness implies mutual respect, affection, and gratitude toward everything that creates or supports life.

Indigenous wisdom matters in today's world of technology, which is making life in industrial societies more convenient but at price that the earth's systems have had to pay. Out of neglect or ignorance, the self-sustaining lands of Indigenous peoples are being destroyed such that not even those who are rooted in Mother Earth are sufficiently protected. In other words, everyone's survival depends on the survival of all life, a principle that has been expressed in many ways.

Indian realities involve more than conceptual understandings because they are inseparably linked to experience and evoke deep feelings out of a sense of connectedness. I've been told that some "words" in a Native language have a deeper meaning that derives both from memory and experience; it is the missing element in many discussions about science, religion, and philosophy. Lovelock (author of the Gaia theory) examined the earth's systems from the viewpoint of a physiologist and reported substantial evidence that the earth is a single living organism. Yet, here again, when Indians refer to Mother Earth, it is personal and experiential, not in terms of a deity or physiology. Renewal is also evident everywhere and need only to be observed in daily life. We two-leggeds were given the necessary faculties to perceive and contemplate the natural world and its hidden realities so that in order to walk accordingly.

The Circle of Power and Healing

Nature works in cycles—periodic, repetitive movements. They exist in the rotating galaxies, the revolving solar systems and planets, the earth's motion, systems and subsystems, down to the atoms, the orbital electrons, and the strings of M-theory. Periodic vibrations are cyclical also because they consist of repetitive motion. The circle symbolizes the periodic movements of nature; it is also an ubiquitous symbol that takes on several forms in Indigenous communities. A pebble thrown into a pool or a lake creates circular waves that radiate outward. If we look carefully at any given point on the water surface, we will notice that it rises and falls at equal intervals.

Our heart has a periodic beat that appears in the form of a sinusoidal wave on a hospital monitor because the beat amplitude is plotted vertically against time along a horizontal axis. If it stops beating, a flat line will appear on the monitor screen. We don't have to think of any of these events as science; they are part of the world in which we live. Each of these natural periodic movements can be represented by a circle. Energy moves through space or other medium in the form of a wave; the higher the amplitude, the greater the energy (except for light). The circle is not a human invention but a design of nature. We often say also that the drum represents the heartbeat of Mother Earth.

The circle is the archetype of periodic motion, balance, and wholeness. Tribal societies have for countless generations harmonized with the cyclical/circular patterns in nature. How many square structures are evident in the natural world? A drop of water falling from the sky or a faucet tends to be spherical.

Animals also follow the natural cycles. Black Elk, holy man of the Oglala Sioux, once said: "Birds make their nests in circles, for theirs is the same religion as ours." "The wind, in its greatest power, whirls. ... You have noticed that everything an Indian does is in a circle, and that is because the Power of the World always works in circles, and everything tries to be round."

The earth wobbles as it moves around the sun in a *precessional* cycle that takes approximately 26,000 years to complete. The "long count" on one of the Mayan calendars, which began in 3114 BC, ended on the winter solstice, December 21, 2012—a date that marked the end of the precessional cycle.

A long-term deviation from the circle would have consequences. It was Western culture that created the square structures in our society. When actions are imposed against the natural order, nature tries to make corrections. The natural disasters often perceived as "acts of God," are they not Earth's attempt to restore balance?

The roundness metaphor from nature tells us how wrong it was to force Indians away from their traditional way of life, to force their psyche into linear structures, and to follow the clock, all of which created pervasive social problems. Instead of finding help in the churches, where help should be available in view of their major role in the multi-generational trauma that afflicts Indian country, Natives have turned to rehabilitation centers and to their own spiritual leaders. For example, the sobriety movement within Indian country is strong. Indigenous models of recovery are effective because they help Indian clients move *back into the circle* of spirituality to recover their Indian identity. The circle is evident in effective approaches to healing, such as the Red Road Approach, the Sacred Tree Curriculum, and White Bison's method, which combines the Medicine Wheel with a version of the Twelve Steps used in addiction programs. In addition to these approaches, psychologists Eduardo and Bonnie Duran are using clinical approaches that take historical trauma into account.

The Medicine Wheel is an ancient teaching and healing system that reflects Creation's cycles and is conceptualized differently among tribes. To be healthy means that all aspects of the human being—spiritual, emotional, physical, mental—are in balance.

Many tribes continue to sustain their way of life and spiritual essence, albeit under changed conditions. Worthy of note are the unselfish prayers of Native people who, despite being extensively marginalized, continue to pray for everyone in the world. In an e-mail that was sent through Native networks, a Native man from Alaska brought an encouraging message from Hopi, Maori, and Stoney elders. "They also want us to know that among the Hopi and Maori there are people who do nothing but pray 24 hours a day, 7 days a week, 365 days a year, every year of their lives. That is all they do. In rotation, they pray around the clock for other people. In this prayer is where they have seen some of these things that are about to happen. The healing that is going to take place, the advice that has been given to us, is 'Seek not to fight evil—do not fight it—let goodness take its place.'"[143]

Native peoples contemplate the moral, ethical, and spiritual aspects of Power in the cosmos. Earth's gifts are meant to be used in a proper, respectful relationship, not as "resources" to be consumed selfishly. The power of nature is not primarily a physical phenomenon to be exploited for human purposes. But we look for healing to come.

The Ultimate Source of Matter

The validity of Einstein's famous equation, $E=mc^2$, which expresses the equivalence of mass and energy, is demonstrated by the creation of new particles during high-energy particle collisions; these are transformations between matter and energy. The equivalence principle is built into the Einstein equations of general relativity because every property of a physical body comprised of mass or energy acts as sources of gravity.

All physical things in the Universe ultimately stem from energy. Particles are transformed into pure energy and in the reverse process, energy can be transformed into other particles. Energy manifests as matter, and matter is created from pure energy. That all things are imbued with spirit and that the Creation is the material manifestation of the spiritual Universe are long-held Native beliefs which are consistent with, but not necessarily implied by, modern physics. Substituting "spirit" for "energy" expresses the Native paradigm.

Atoms, and consequently all objects, consist of three key material particles—the proton, the neutron, and the electron—and three of the four fundamental forces of nature—the strong force, the electroweak force, and the electrostatic force. At larger scales it becomes evident that objects attract each other through gravity, which keeps the planets, solar systems, stars, and galaxies organized and in cyclical motion. Gravity cannot exist without matter (or energy).

Intelligent Design

"Intelligent Design" advocates argue their case within the empirical framework of Western science which, in my opinion, is a fatal mistake. Intelligence is in the creation, not only its cause. When the controversy began, parents were concerned because evolutionary theory was taught as a fact which conflicted with their convictions. Part of the problem was, and still is, that "evolution" has two meanings: does it mean that new species evolve or that changes occur only within the same species, the latter of which agrees with the first chapter of Genesis. But because the former, broader meaning of evolution is the one taught in the schools and the one most people assume, many Bible-believing people altogether avoid using the word "evolution."

Neither side of the debate is without error yet each side stands firm in its position and, needless to say, the controversy shows no signs of abating. Certain beliefs may be correct but, without empirical proof, they cannot be considered science under the Western paradigm. Parents, scientists, and others are polarized and emotions are high, judging from the way people rally when a court or a board decision is imminent; sadly, when a court of school decision is made, there are always winners and losers.

America has deviated far from its original spiritual vision that once accommodated references to the sacred trust between the nation (actually, all humankind) and its Creator, a trend that has added more empirical requirements to the meaning of science. Intelligent Design advocates apparently do not recognize the impact of this trend and are advancing ID as a parallel theory within a paradigm that requires adherence to material causes as proof of nature's complexity. They argue that Intelligent Design is science because it represents an unrestricted search for the truth about nature based on reasoning from physical evidence. Claiming that intelligence is evidence of complexity is equivalent to appealing to a higher power, or God, according to the judge, and therefore not a science-based argument.[144] ID advocates countered with the argument that an unrestricted search for the truth about nature, based on reasoning from physical evidence, is science.

It seems more reasonable to say that intelligence is evident in nature. This was Einstein's attitude who was humbled by the superior intelligence manifested in the laws of the Universe—a personal conviction that he omitted from his theories.

It is the attitude or the approach to a theory or hypothesis that is philosophical or religious, not the theory itself. Einstein's emotional response to a harmonious Universe, for example, was the motivation behind his discoveries, yet his belief did not become a part of his physics. His approach to science converges with American Indian metaphysics insofar as we are awed by nature's mysteries, which inspire us to investigate, but we must be meticulous in our attempts to ascertain what nature is saying when she reveals a secret.

Physical theories are generally more complex than the general public can understand but scientists should also be honest about their own assumptions, beliefs, and biases. Producing evidence that evolution is possible, or actually occurring now, does not prove that life actually originated that way; in this sense evolution is a hopeful but unproved theory. By taking this approach, the school children will not have to listen only to one-sided arguments in favor of evolution; they will also know that many people, including people in science, are willing to say that they also believe what the Bible teaches. There is no need to feel defeated in a free country.

Native people are not offended when others express their faith in public; the beliefs of others are respected; in fact, a leader will usually say, "pray in your own way," when a public prayer is about to be expressed. Prayers to the Creator occur at virtually every Indian gathering; spirituality is an essential part of tribal life. Religious controversies have occurred in the tribal communities when they were brought in from outside. This has been my experience.

The Seventh Generation

The U.S. Constitution states that one of its purposes is to "secure the Blessings of Liberty to ourselves and our Posterity" but it does not guarantee explicit rights to clean (uncontaminated) land, air, and water. In traditional Indigenous thought, no one owns the land, the water, or the air—these are to be shared in common. Long before the Constitution was conceived, the Seventh Generation Principle had been at the center of sovereign tribal nations' decision-making role. Simply stated, we in the current generation should make decisions about how we live today based on how our decisions will affect the future seven generations.

The following statement appears in the Constitution of the Iroquois Nations: The Iroquois Great Binding Law, *Gayanashagowa*:[145]

> In all of your deliberations in the Confederate Council, in your efforts at law making, in all your official acts, self interest shall be cast into oblivion. Cast not over your shoulder behind you the warnings of the nephews and nieces should they chide you for any error or wrong you may do, but return to the way of the Great Law which is just and right. Look and listen for the welfare of the whole people and have always in view not only the present but also the coming generations, even those whose faces are yet beneath the surface of the ground -- the unborn of the future Nation.

Oren Lyons, Chief of the Onondaga Nation, has also written the following statement: "We are looking ahead, as is one of the first mandates given us as chiefs, to make sure and to make every decision that we make relate to the welfare and well-being of the seventh generation to come. . . ." "What about the seventh generation? Where are you taking them? What will they have?"[146]

In 1977, Winona LaDuke (Anishinaabe) was among the proponents of a Seventh Generation Amendment to the U.S. Constitution, also called the Common Property Amendment, that does not clash with any other constitutional amendments (including private property rights). It reads: "The right of citizens of the United States to use and enjoy air, water, sunlight and other renewable resources determined by the Congress to be common property shall not be impaired, nor shall such use impair their availability for the use of future generations."[147]

I do not know the current status of this proposal to amend the Constitution but it seems to be an active campaign.

Tell a Story

Events at the quantum scale cannot be described in terms of objects, like rocks and other things, that move under the influence of forces because there are no objects; it is not a mechanistic world of cause and effect. There are events and relationships between them, such as what happens in a story. "The world is a history of processes. Motion and change are primary," says physicist Smolin.[148] "When we are dealing with processes instead of objects, to ask how something is, is an illusion."

Smolin offers the following solution for describing quantum processes: tell a story. The question of what causes events and why they occur involves the flow of information between them and any event is typically the result of several influences, not just one. Considering what Rael, Smolin, and others have said, the quantum world of processes could involve energy, spirit, or mind.

I have listened to elders tell stories and, over the years, my heart has become more tender and my mind more inquisitive about what lies beyond our senses as a direct result of the experience. Stories perpetuate oral tradition and language. They transport us to another world, to a different state of consciousness where everything important about life seems to come together in another dimension and time has no meaning, or a different meaning, making events somehow seem more real. Stories sharpen our listening skills, transmit the tribe's history and knowledge of its origin to future generations, teach moral lessons, and tell why things are the way they are. They teach respect and clarify concepts. The people learn who they really are and their purpose on Earth, and why they call themselves by their name. Children participate in the culture since birth and learn to use their imagination as they listen to tribal myths that teach about the real world. Dreams are also a channel to knowledge. The world of stories and dreams is an unbounded metaphysical world. Much of this richness has been lost due to colonization.

Embracing the Sacred Web of Life

As I have tried to describe, Indigenous knowledge includes tribal philosophy in its perception of reality as it relates to the totality of human experience. It has qualities that are direly needed by Western societies but it is not the same kind of knowledge as Eurocentric science. Indigenous knowledge varies from tribe to tribe because it is of a local and oral nature. However, certain tribal concepts and principles are shared by many tribes and possess universal value.

While eminent physicists refer to the Universe as a web of relationships due to its objective wholeness, Indigenous peoples know it experientially and more than a physical concept; they contemplate and relate to a living Universe as a social reality, not an object to be studied, and humans are a part of it.

The cosmos as web of relationships is inferred from modern physics and stated explicitly by physicists conducting research in cosmology. But can scientists embrace the *sacred* web of life without compromising the integrity of their discipline? I believe they can, not by appending it to science but by reassessing and adjusting the approach that most physicists and other researchers use in their endeavors to learn about nature. Implementing better strategies and changing the predominant perception of nature from something to be exploited to something to be respected should not be a problem, considering that holistic strategies and attitudes are already being proposed under various names such as "deep ecology" and "environmental science," which, ironically, promote some of the very same principles that Native peoples have long practiced.

These organizations and academic programs do not perceive the cosmos as something sacred but this is what is lacking in the current science paradigm. A consciousness of the sacred does not preclude the need for testable theories. In fact, it is likely to improve the integrity of scientific endeavors while preserving and increasing our potential for survival. We will live only if *all our relatives* in the web of life are allowed to live.

Appendix

THE DOCTRINE OF DISCOVERY

American history did not happen by chance; events were driven by a common ideology. Documented history is physically recorded but the unseen world of intent and spirit is difficult to perceive. History has been influenced by insidious doctrines that have been used to justify United States expansionism and wars of conquest. An acquaintance with these doctrines will help Americans understand their own history, how to better relate to tribal people, and how to prevent history from repeating itself. Since the arrival of Europeans to the northeastern seaboard (for the Pueblos of New Mexico it was much earlier), American Indians and immigrant Americans have shared a common history.

The voyages of Christopher Columbus precipitated a new era of discovery, making it necessary for European powers to formulate juridical standards for obtaining title to newly discovered lands. These standards originated with the Pope and were eventually incorporated into U.S. law. What came to be known as the Doctrine of Discovery, also called the Right of Christian Discovery, represents the primary tenets of international law. It developed from the notion formulated about forty years before Columbus set sail and is still in force despite efforts in recent years to have it repealed. It asserts that Christian nations have a divine right to ownership of lands that they would eventually "discover," and that they also have authority over the (Indigenous) peoples inhabiting those lands.

In "Five Hundred Years of Injustice: The Legacy of Fifteenth Century Religious Prejudice," Steve Newcomb describes two papal documents. The first one was issued in 1452, forty years before Columbus' historic voyage of 1492. Pope Nicholas V declared war against all non-Christian nations throughout the world, declaring them enemies of the Catholic faith. He promoted the conquest, capture, and enslavement of the Native inhabitants, and the seizure of their possessions and properties. This authorization was in place at the time that Columbus set sail. When Columbus returned from his first voyage, Pope Alexander VI issued another papal document in 1493 that "granted" Spain the right to conquer the lands that Columbus had already found and lands that have yet been undiscovered. The acts of genocide and brutality against Native people were sanctioned by these documents.

The Discovery Doctrine as it existed at the time of Columbus may be easier to understand in the following summary, which I derived from Ward Churchill's book, *Struggling for the Land*:[149]

1. Land claims are valid *only* for Christian emissaries. This rule applies to all lands discovered.

2. If the land is uninhabited, it belongs to the discoverer.

3. If the land is inhabited:
   the Native people's inherent rights must be honored;
   only the discoverer can acquire land;
   the discoverer is obligated to evangelize;
   armed force is unjustified without the Natives' consent.

The Condor and the Eagle

4. Exceptions: If any one of the cases below is true, the discoverer has the "legal" right to make "just war":
    a) the Natives refuse to trade;
    b) the Natives reject (refuse) the missionaries;
    c) the Natives attack the discoverers.

The doctrine was incorporated into U.S. law in two key court decisions in which Chief Justice John Marshall wrote the U.S. Supreme Court's opinion. The first of these is *Johnson v. M'Intosh* (1823) he claimed that the "heathen" Indians had lost their right to full sovereignty and that the United States, after winning its independence from Great Britain, had acquired the power of dominion over Indian lands. He also argued that the "inferior" character and religion of American Indians made them a people over whom the "superior genius" of Europe should rule. He wrote: "The character and religion of ... (American Indian) inhabitants afforded an apology for considering them as a people over whom the superior genius of Europe might claim an ascendancy."[150]

In *Cherokee* Nation *v. Georgia* and *Worcester v. Georgia*, Chief Justice Marshall proceeded on the basis of his unfounded prior conclusions (*McIntosh*) to argue that American Indian peoples comprised nations domestic to and dependent upon the United States. They occupied a status of "quasi-sovereignty," he asserted, being sovereign enough to engage in treaty-making with the U.S. (for purposes of conveying legal title to their lands), but not sovereign enough to manage their other affairs as fully independent political entities."[151]

In case it is not yet clear to the reader, the Doctrine of Discovery, U.S. federal Indian law, and the notion of "Manifest Destiny" *are neither ethical nor legal;* they are *beliefs* based on religious and racial prejudice. Divine Providence is also cited as the primary cause of historical events that favored the settlers over the Indians, including the dominant view being taught to children that Indian villages were providentially decimated by disease in order to make room for the immigrants.

In "The Discovery Doctrine, the Tribes, and the Truth," Jack Utter states that the Discovery Doctrine has formed the legal basis of the transfer of 97 percent of America from Indian to non-Indian ownership. Regarding the superiority of one people over another, he refers to a "revealing" 1996 book by Moffitt and Sebastian who traced the roots of the Discovery Doctrine to Aristotle. They state that Aristotle advanced a dangerous thesis about "Barbarians" who need to be ruled by their more rational "betters" (the Greeks). Aristotle stated that "barbarian and slave are, by nature, identical" and that "in this world there are but two kinds of men, that kind which naturally rules, and that which naturally is ruled..."

Utter further states that "Spanish secular and church authorities and legal scholars, recognized today for being the first to apply the Discovery Doctrine to the Western Hemisphere and for having developed foundations of modern federal Indian law, customarily relied on Aristotle's writings (and those of the church) to justify their actions and adventures in the 'New World.' Further, Aristotle's ancient writings provided a philosophical or cultural license for the Spanish and other Europeans, to combine with their religious license, in justifying the displacement, dispossession, and even destruction of other perceived inferior peoples, including the Native peoples of North and South America." Utter notes that, although Aristotle lived before the Christian era, Christian scholars and church leaders later combined cultural and Christian differences about people into a new worldview in seeking justification for religious superiority. Later in the same article, he states: "The Discovery Doctrine was therefore applied to any race of people whose religion and culture varied from a primary Christian

European norm. That is why the doctrine (which is ultimately legal fiction) is often referred to as a 'culturally racist' precept, as opposed to a biologically racist one..."[152]

Reviewers' Comments

Several knowledgeable individuals—Native American culture bearers, scholars, experts from diverse academic disciplines—reviewed the manuscript for this book and submitted their comments, some of which are excerpted at the beginning of this book. It would be a great disservice to everyone if any part of their valuable statements were omitted. For those whose statements were excerpted, below are their comments *in full*.

~~~~~~~~~~~~~~~

Dr. Ray Barnhardt
Professor of Cross-Cultural Studies
Indigenous Knowledge Systems
University of Alaska Fairbanks

"Phillip Duran's treatise, *The Condor and the Eagle: Uniting Heart and Mind in Search of a New Science Worldview*, provides a comprehensive critique and analysis of many of the complex philosophical issues that come into play at the intersection of indigenous ways of knowing and Western science. He has been able to not only engage in a highly reflexive inquiry into some of the most salient issues facing indigenous societies, but to also offer some relevant and practical examples of how those issues might be addressed in contemporary cross-cultural contexts. This is no small task, but he has succeeded in delineating and elaborating on the dialectics of interaction across diverse scientific and cultural arenas in a way that expands our understanding and opens up many new questions and avenues for inquiry. Dr. Duran reaches into the depths of the philosophy of science to offer us new insights that are of benefit to the whole of human existence.

The Condor and the Eagle

"Dr. Duran's treatise provides insightful analysis of the deep and complex ontological and epistemological issues that come into play when quite different systems of thought converge. In critiquing the evolution of diverse thought systems, he draws upon vast bodies of historical knowledge to demonstrate how such complex issues can be made coherent and intelligible. While some of the theoretical constructs and frames of reference outlined in the early chapters were put to the praxis test in later chapters, Dr. Duran is in fact quite consistent in drawing together a wide range of examples that provide ontological context for indigenous being, knowing and doing that serves as the focal point for much of his analysis. This is a piece of work that clearly extends the boundaries of our understanding of the dynamics at the interface of diverse epistemological traditions, and he effectively makes the case for a more critical and culturally-based approach to the practice of science in the indigenous world."

~~~~~~~~~~~~~~~

Dr. Vivian Delgado (Yaqui)
Professor of Indigenous (Indian) Studies
Bemidji State University
Two Cognates: Native American Philosophy and Higher Education

"Indigenous Studies can be a very dark subject to study, so I have always encouraged my colleagues to take care of themselves. Throughout this semester I have read the work of Phil Duran and by its end I knew this was part of the 'taking care of myself' that I was preaching to others because our intellect (our oldest part of self) also needs reassurance and care. Duran's words took care of numerous intellectual and academic issues that have alarmed me during my lifetime and carry significant meaning in my work.

The Condor and the Eagle's stated intent—that its message 'should interest and challenge all readers'—is an understatement; I was humbled by its message. What I have found is a bold new area of science that recognizes hidden realities revealed by modern physics. Furthermore, Indigenous American physics is very fitting for Indigenous Philosophy as part of the realignment of science and who we are becoming in the modern world.

Indigenous American physics is about modern physics, Indigenous knowledge and Western science. Duran has accumulated a lifetime of wisdom from which he conveys his knowledge of Indigenous intelligence and philosophy pertaining to 'the natural world to Western scientific minds and explains relevant scientific concepts in terms that are familiar to Native peoples.' I find the smaller and fine explanations will be appealing to Indigenous students by offering them an important part of our missing identity that connects our past with our modern being. Indigenous readers will be amazed that the quantum's divergent behaviors accommodate Indigenous concepts that 'all things ultimately stem from spirit and that the Universe is alive.'

Duran's book offers a new paradigm in which expressions, knowledge, and social justice realities are free of religion. All readers need to know that 'Western values have contradicted the very holism and authoritative principles inherent in nature and revealed by modern physics.' This is a reaffirmation that Indigenous people have always held; that the proper way to live is linked to the understanding of and respect for nature.

Among Indigenous North Americans we need more discussion about 'our living in a world of illusions, unaware of the unseen world of reality that supports the whole universe.' We need to discuss the lack of connection between knowledge and morality and the strong connection between knowledge and power and the political influence of wealth because; 'the thousands of Indigenous nations worldwide outnumber the industrial societies formed from the conquest of tribal people.'

May all readers be blessed with Duran's awareness 'of a universe that is endowed with its own consciousness, that is intelligent, authoritative, and harmonious – the manifestations of a spiritual reality from which the Creator made all things and to whom we are responsible.' What we have here in this book is a spiritual, scientific prayer that is years ahead of its time and will surely lead the way to a moral, spiritual and intellectual consciousness."

~~~~~~~~~~~~~

Dr. Leroy Little Bear
Blood Tribe of the Blackfoot Confederacy
Former Director, American Indian Program at Harvard University
Professor Emeritus of Native Studies at the University of
Lethbridge

"Einstein said that the business of science is 'reality'. But, whose reality? To a large extent, what is commonly referred to as 'science' has largely been monopolized by Western cultures. The view of science, from a Western perspective, is one of measurement based on mathematics. But there are other approaches to the human search for reality. One of those other approaches is the search for reality based on North American Indian paradigms. These paradigms include but not limited to concepts of constant flux, all of creation consisting of energy waves, everything being animate, everything being interrelated, reality requiring renewal, land as a major referent, and language as a repository for knowledge. When one looks through these paradigmatic lenses, one sees a very different version of 'reality'.

"The seeing of a very different version of 'reality' is one reason why North American Indians are forever explaining themselves to Westerners. The explaining echoes what the Ohio Humanities Council says about the humanities in academia: "The humanities are the stories, the ideas, and the words that help us make sense of our lives and our world. The humanities introduce us to people we have never met, places we have never visited, and ideas that may have never crossed our minds." In other words, Native Americans become wonderful and great storytellers because they introduce to Westerners ideas that have never crossed their minds.

"In my many nomadic wonderings, physically over 'turtle island' and academically, I am always in search of Native Americans who have a special interest in science, especially in quantum physics. Why quantum physics? The answer is that many of the concepts in quantum physics are very similar to concepts that arise out of Native American paradigms. For a number of years I was part of a small group of Native Americans that got together with Western physicists to explore similarities and differences between quantum physics and Native American science. These gatherings we referred to as 'science dialogues.' It was at one of these gatherings in Albuquerque, New Mexico that I first met Phil Duran. I became very awe-struck by his knowledge of physics and how he was able to transcend the boundary between Western and Native American scientific worldviews. This book, "The Condor and the Eagle" is a wonderful exercise in 'boundary crossings'; it is a wonderful exercise in 'storytelling'.

"Phil Duran is a great storyteller who, as a physicist, was very brave to cross the boundary between two scientific worlds.

He is a true warrior, who is brave enough to cross the boundary into the Western physics world to explicate the Native American scientific way. I am twice blessed: I found another Native American scientist steeped in quantum physics and "THE CONDOR AND THE EAGLE' is a wonderful story about boundary crossings between Western and Native American sciences. It is a 'must' read for scientists, both Western and Native American."

References

[1] Hawken, Paul. *Blessed Unrest: How the Largest Movement in the World Came into Being and Why No One Saw It Coming*. New York: Penguin Group, 2007.

[2] Einstein, Albert. *The Meaning of Relativity*. Princeton University Press, 1988. Introduction by Brian Greene, 2005, p.vii.

[3] Deloria, Vine. *Evolution, Creationism, and Other Modern Myths*. Golden, Colorado: Fulcrum Publishing, 2002, p. 167.

[4] *Oren Lyons: The Faithkeeper*. The Moyers Collection [VHS] (1991).

[5] Commanda, William. "Seven Prophets, Seven Fires," in *Profiles in Wisdom: Native Elders Speak about the Earth*. McFadden, Steve, ed. Santa Fe: Bear & Company, 1991.

[6] Brooks, Drex. *Sweet Medicine: Sites of Indian Massacres, Battlefields, and Treaties*. 1995. Albuquerque: University of New Mexico Press.

[7] Lyons, Oren. *Oren Lyons: The Faithkeeper*. The Moyers Collection [VHS] (1991). Also, in *A Story Waiting to Pierce You: Mongolia, Tibet and the Destiny of the Western World*. Point Reyes, California: The Golden Sufi Center, 2010, p. 76 and note #30, Peter Kingsley explains that the tradition of the bundle of arrows can be traced to ancient encounters between Native Americans and Mongolia, Siberia, and Central Asia.

[8] Whitehead, Alfred N. *Science and the Modern World*. New York: The Free Press, 1997, p. 7.

[9] Feulner, Edwin, "Roots of Conservative Thought from Burke to Kirk," *First Principles Series* 19 (July 9, 2008), Washington, DC: The Heritage Foundation.

[10] Kingsley, Peter. "Original Instructions: An Interview with Peter Kingsley," in *Quest*, Volume 99, Number 3, Summer 2011.

[11] Lemay, Connie. "A Brief History of American Indian Military Service," in *Indian Country Today Media Network*, May 28, 2012, URL = <http://indiancountrytodaymedianetwork.com/article/a-brief-history-of-american-indian-military-service-115318>.

[12] E-mail excerpt from Michael Moore's Christmas eve 2012 letter to his distribution list.

[13] Quoted by Hawken, Paul. *Blessed Unrest: How the Largest Movement in the World Came into Being and Why No One Saw It Coming*. New York: Penguin Group, 2007, p. 87.

[14] Curd, Patricia, "Presocratic Philosophy", *The Stanford Encyclopedia of Philosophy* (Spring 2012 Edition), Edward N. Zalta (ed.), URL = <http://plato.stanford.edu/archives/spr2012/entries/presocratics/>.

[15] Aikenhead, Glen. "Integrating Western and Aboriginal Sciences," University of Saskatchewan, 2002, URL = <www.usask.ca/education/people/aikenhead/index.htm>.

[16] Indian Nations at Risk Task Force. (1991, October). *Indian nations at Risk: An educational strategy for action* (Final report of the Indian Nations at Risk Task Force). Washington, DC: U.S. Department of Education.

[17] Schaefer, Carol. *Grandmothers Counsel the World: Women Elders Offer Their Vision for Our Planet*. Foreword by Winona LaDuke. Boston: Trumpeter Books, 2006. Kindle Edition, p. 1. Their website is at URL = <http://www.grandmotherscouncil.org>.

[18] Ereira, Alan. *The Elder Brothers' Warning*. Tairona Heritage Trust, 2009. The book has been re-published by Tairona Heritage Trust (www.taironatrust.org), a British-based NGO that has been working in behalf of the Gonavindua Tairona (the Kogi culture).

[19] The Community Adaptation and Sustainable Livelihoods (CASL) website is at URL = <www.iisd.org/casl/projects/inuitobs.htm>.

[20] The Alaska Native Knowledge Network (ANKN) website is at URL= <http://www.ankn.uaf.edu/>.

[21] Akwesasne notes, ed. *A Basic Call to Consciousness: The Haudenosaunee Address to the Western World*. Summertown, TN: Native Voices. (New edition, June 2005). This edition includes a detailed account of the journey to Geneva and other articles, as well as the original 1977 address, which was published in 1978. The entire 1977 address may be read online at http://www.ratical.com/many_worlds/6Nations/BasicCtC.html.

[22] Aikenhead, Glen and Michell, Herman. *Bridging Cultures: Indigenous and Scientific Ways of Knowing Nature*. Toronto: Pearson Canada, Inc., 2011, p. 4.

[23] Lovelock. *Gaia: A New Look at Life on Earth*. Oxford: University Press, 2000.

[24] Capra, Fritof. *The Tao of Physics*. Boston: Shambala Publications, Inc., 1983.

[25] Kingsley, Peter, Summer 2011, *op. cit.*

[26] Kingsley, Peter. *Ancient Philosophy, Mystery, and Magic: Empedocles and Pythagorean Tradition*. Oxford: Oxford University Press, 1995.

[27] Quoted by Peter Kingsley in *Ancient Philosophy, Mystery, and Magic: Empedocles and Pythagorean Tradition*, Oxford University Press, 1995, pp. 229-230.

[28] Lindberg, David C. *The Beginnings of Western Science*. University of Chicago Press, 2007, p.27.

[29] Momaday, N. Scott. "An American Land Ethic," in *The Man Made of Words: Essays, Stories, and Passages*. New York: St. Martin's Press, 1998.

[30] Einstein, Albert. *Relativity: The Special and the General Theory*. New York: Crown Publishers, Inc., 1961, p. 2.

[31] Lindberg, David C. *The Beginnings of Western Science*. University of Chicago Press, 2007, p. 82.

[32] Men, Hunbatz. *Secrets of Mayan Science/Religion*. Santa Fe: Bear & Company. 1990, p. 25.

[33] Einstein, Albert. *Albert Einstein: Out of My Later Years*. New York: Winds Books. 1956, p. 62.

[34] *Ibid.*, p. 98.

[35] Hebrews 11:1-3 (NIV)

[36] Deloria, Vine, Jr. *The World We Used to Live In: Remembering the Powers of the Medicine Men*. Golden, Colorado: Fulcrum Publishing, 2006, p. xvii.

[37] Isaacson, Walter. *Einstein: His Life and Universe*. New York: Simon and Schuster, 2007, p. 388.

[38] *Ibid.*

[39] Mih, Walter. *La Fascinante Vida y Teoría de Albert Einstein, (The Fascinating Life and Theory of Albert Einstein)*. Francisco Manzo-Robledo (translator), Mexico: Libros para Todos, S.A, pp. 9-13.

[40] *A Basic Call to Consciousness*, op. cit.

[41] LaViolette, Paul A. *Genesis of the Cosmos*. Rochester, Vermont: Bear & Company, 2004, p. 9.

[42] *Ibid.*, p. 3

[43] Deloria, Vine. *Evolution, Creationism, and Other Modern Myths*. Golden, Colorado: Fulcrum Publishing, 2002, p. 164.

[44] *Ibid.*, p. 166. Deloria's source for these dates is Peter Tompkins, *Mysteries of the Mexican Pyramids*.

[45] Accessed at URL = <www.antiwar.com/ips/marquez.php?articleid=4859> on February 4, 2013.

[46] Keoke, Emory Dean, and Porterfield, Kay Marie. *Encyclopedia of American Indian Contributions to the World: 15,000 Years of Inventions and Innovations*, Facts on File Library of American History, 2002.

[47] Aczel, Amir. 2001. *Entanglement*. New York: Penguin Books, p. ix. Amir Aczel is the author of *God's Equation* and other works and the critically acclaimed author of *Fermat's Last Theorem*.

[48] Ibid., Kakalios, James. *The Amazing Story of Quantum Mechanics*. New York: Gotham Books, 2010, p. 197.

[49] *Ibid.*, p. 215.

[50] Petersen, A. "The Philosophy of Niels Bohr," *Bulletin of the Atomic Scientists*. September 1963. This statement appears in more than one place. See, for example, Wikipedia article at URL = <http://en.wikipedia.org/wiki/Quantum_mechanics,_philosophy_and_co ntroversy>.

[51] Kafatos, M. and Nadeau, R. *The Conscious Universe: Parts and Wholes in Modern Physical Reality*. New York: Springer, 2000, p. 34.

[52] Kafatos, M. and Nadeau, R. 2000, *op. cit.*, p.43

[53] Farmelo, Graham (2009-08-25). *The Strangest Man: The Hidden Life of Paul Dirac, Mystic of the Atom* (Kindle Locations 2648-2650). Perseus Books Group. Kindle Edition.

[54] Elgin, Duane. "Scientific Evidence of a Living Conscious Universe," *Dream Manifesto*, June 8, 2008, URL = <http://www.dreammanifesto.com/scientific-evidence-living-consciousness-universe.html>. This article is a commentary on the book, *Mind before Matter: Vision of a New Science of Consciousness* (2007), co-edited by Trish Pfeiffer, John E. Mack, and Paul Devereux.

[55] Elgin, Duane. "We Live in a Living Universe," *Working with Oneness, Working with Oneness,* June 8, 2008, URL = <http://www.workingwithoneness.org/articles/we-live-living-universe>. The quote is from Freeman Dyson's book, *Infinite in all Directions*.

[56] Chopra, Deepak. "Science and spirituality should be friends," *My Take CNN Belief Blog*, February 15, 2011, URL = <http://religion.blogs.cnn.com/2011/02/15/my-take-science-and-spirituality-should-be-friends>.

[57] Josephson, B. and Rubik, B. "The Challenge of Consciousness Research," accessed on 1/14/2013 at URL = <http://www.tcm.phy.cam.ac.uk/~bdj10/mm/articles/athens.txt>.

58 Einstein, Albert. *The Meaning of Relativity*. Princeton University Press, 1988. In the book's 2005 Introduction on p. vii, Brian Greene states: "Nevertheless, even though relativity has been systematized mathematically, the vast majority of physicists would say that they still don't 'feel relativity in their bones.' I, for one, know how easy it is to slip into Newtonian thinking…"

59 Hawking, S. and Mlodinow, Leonard. *The Grand Design*. Bantam Books, 2010, p. 7.

60 Rael, Joseph. *In the House of Shattering Light*. Special Edition. 2002, back cover.

61 Davis, Wade (2011-05-29). *The Wayfinders: Why Ancient Wisdom Matters in the Modern World* (CBC Massey Lecture). Perseus Books Group. Kindle Edition, location 1653.

62 Hawking, S. and Mlodinow, Leonard. *The Grand Design*. Bantam Books, 2010, the Rule of Law chapter.

63 *Ibid.*

64 Isaacson, Walter, *Ibid.*, pp. 124-125.

65 Einstein, Albert. *Relativity: The Special and the General Theory*. New York: Crown Publishers, Inc., 1961, p. 13. The statement cited here is equivalent to the one quoted.

66 Bohm, David. *The Special Theory of Relativity*. London and New York: Routledge, 1996.

67 *Ibid.*, pp.91-94.

68 Rael, Joseph. *Ceremonies of the Living Spirit*. Tulsa/San Francisco: Council Oaks Books, 1998, p. 31.

69 Hunbatz, *Ibid.*, Men, p. 24.

70 Bohm, David. *The Special Theory of Relativity*, *Op. cit.*

71 Lindberg, David C. *The Beginnings of Western Science*. University of Chicago Press, 2007, p. 251.

72 Sando, Joe S. *Nee Hemish: A History of Jemez Pueblo*. Santa Fe: Clear Light Publishing, p.ix

73 Kafatos, M. and Nadeau, R. *The Conscious Universe: Part and Whole in Modern Physical Theory*. New York: Springer-Verlag, 1990, pp. 4-5.

74 Nichols, Edwin J. "The Philosophical Aspects of Cultural Differences." Cultural Bias Methods of Analysis Conference speech on DVD, Men and Women of Respect Historical Society, URL = www.MenofRespect.com.

[75] Capra, Fritjof. *The Tao of Physics, op. cit.,* chapter one.

[76] Johnson, P. 2000. *The Wedge of Truth.* Downers Grove, Illinois: InterVarsity Press, 2000, p. 154.

[77] Ibid., po. 160.

[78] Davis, Wade. *The Wayfinders: Why Ancient Wisdom Matters in the Modern World* (CBC Massey Lecture). Perseus Books Group. Kindle Edition, loc. 1837, 2011.

[79] Johnson, P. 2000. *The Wedge of Truth.* Downers Grove, Illinois: InterVarsity Press, 2000, pp. 154, 160-162.

[80] Aikenhead, G. "Integrating Western and Aboriginal Sciences: Cross-Cultural Science Teaching," in *Research in Science Education,* 2001, vol. 31, no. 3, pp. 337-355.

[81] Lyons, Oren. 1984. "Our Mother Earth," *Parabola,* Volume 7, No. 1, Winter, 91-93.Nelson, Melissa K., ed. *Original Instructions.* Rochester: Bear & Company, 2008, p. xxiii.

[82] Bopp, Judie. *The Sacred Tree.* Twin Lakes, WI: Lotus Press, 1984.

[83] Acts 17:26-27 (NIV)

[84] Psalms 33:16-17 (NIV)

[85] Napoleon, Harold. *Yuuyaraq: The Way of the Human Being.* Fairbanks: Alaska Native Knowledge Network, 1996, p. 4.

[86] Davis, Wade (2011-05-29). *The Wayfinders: Why Ancient Wisdom Matters in the Modern World* (CBC Massey Lecture). Perseus Books Group. Kindle Edition, locations 595-688.

[87] Liu, Jodie. "Conservation Refugees" in Zócalo Public Square, July 8, 2009, URL = http://www.zocalopublicsquare.org/2009/07/08/book-review-conservation-refugees/book-reviews.

[88] The *Encyclopedia of American Indian Contributions to the World* (Keoke and Porterfield, 2002) contains more than 450 entries.

[89] Akwesasne notes, ed. *A Basic Call to Consciousness: The Haudenosaunee Address to the Western World.* Summertown, TN: Native Voices, June 2005, p.90. This publications contains the entire 1977 message (consisting of three position papers) which was published in 1978. The entire 1977 address is available on the Internet at URL = <www.ratical.com/many_worlds/6Nations/BasicCtC.html>.

[90] Ereira, Alan. *From the Heart of the World: The Elder Brothers' Warning* [VHS], 1998. An urgent message to the "younger brother" (*i.e.,* the industrialized nations) that "unless we work together, the world will die."

[91] "Inuit Observations on Climate Change," Community Adaptation and Sustainable Livelihoods. URL = < www.iisd.org/casl/projects/inuitobs.htm>.

[92] Tex Hall, president of the National Congress of American Indians, speaking to the Native American Fish and Wildlife Society, Anchorage, Alaska, 2001.

[93] Akwesasne Notes, *Op. cit.*

[94] Link to the ANSC website at URL = <www.nativescience.org/html/traditional_knowledge.html>.

[95] Kawagley, Angayuqaq O. and Barnhardt, Ray. "Education Indigenous to Place: Western Science Meets Native Reality," URL = <www.ankn.uaf.edu/Curriculum/Articles/BarnhardtKawagley/EIP.html> .

[96] Deloria, Vine, Jr. and Wildcat, Daniel R.. *Power and Place. Power and Place: Indian Education in America.* Golden, Colorado: Fulcrum Resources, 2001, p.2.

[97] Aikenhead, *op. cit.*, p. 65.

[98] Keoke, Emory Dean, and Porterfield, Kay Marie. *Encyclopedia of American Indian Contributions to the World: 15,000 Years of Inventions and Innovations*, Facts on File Library of American History, 2002. A shorter edition by the same title but for younger readers was published in 2005.

[99] Phillips, John L. "A Tribal College Land Grant Perspective: Changing the Conversation." Journal of American Indian Education. Volume 42, Issue 1. 2003.

[100] Bohm, David. *Wholeness and the Implicate Order.* New York: Routledge, 1980.

[101] Men, Hunbatz, *loc. cit.*.

[102] Rael, Joseph. *In the House of Shattering Light.* Special Edition. 2002, p. 128.

[103] Aikenhead, Glen and Michell, Herman. *Bridging Cultures: Indigenous and Scientific Ways of Knowing nature.* Toronto: Pearson Canada, Inc., 2011, p. 65.

[104] Hurtado, Lorenzo Muelas. "Access to the Resources of Biodiversity and Indigenous Peoples," Edmonds Institute occasional paper. Bogota, Colombia, 1998, URL = < www.edmonds-institute.org/muelaseng.html>.

[105] Armstrong, Jeanette. "Sharing One Skin: the Okanagan Community," in *The Case Against Global Economy*, Jerry Mander and Edward Goldsmith (eds.), San Francisco: Sierra Club Books, 1996, pp. 460-470.

[106] Napoleon, Harold. *Yuuyaraq: The Way of the Human Being*. Fairbanks: Alaska Native Knowledge Network, 1996.

[107] Lyons, Oren. "Our Mother Earth," *Parabola,* Volume 7, No. 1, Winter, 1984, pp. 91-93.

[108] The Mind-Matter Unification Project directed by theoretical physicist and Nobel laureate Brian Josephson is an example. URL = <www.tcm.phy.cam.ac.uk/~bdj10/>.

[109] Capra, F. *The Tao of Physics*. Boston: Shambala Publications, Inc., 1985, pp. 17-25.

[110] Deloria, Vine. *Evolution, Creationism, and Other Modern Myths*. p. 164

[111] Barbour, Ian G.. *When Science Meets Religion*. New York: HarperCollins, 2000, p.34.

[112] Alaska Native Science Commission. "What is Traditional Knowledge?" <http://www.nativescience.org/html/traditional_knowledge.html>.

[113] Hurtado, Lorenzo Muelas. "Access to the Resources of Biodiversity and Indigenous Peoples," Edmonds Institute occasional paper. Bogota, Colombia, 1998, URL = < www.edmonds-institute.org/muelaseng.html>.

[114] Men, Hunbatz, *loc. cit.*

[115] Lindberg, David C. *The Beginnings of Western Science*. University of Chicago Press, 2007.

[116] Capra, Fritjof. *The Tao of Physics, 20.*

[117] Kingsley, 2011, *op. cit.*

[118] Lindberg, David C. *The Beginnings of Western Science*. University of Chicago Press, 2007, p. 39.

[119] Lindberg, David C. *The Beginnings of Western Science*. University of Chicago Press, 2007, p. 139.

[120] Lindberg, David C. *Op. cit.,* p. 219.

[121] Matthew 25: 35-36, 40 (NIV)

[122] Lindberg, David C. *Op. cit.,* pp. 163-193

[123] Whitehead, Alfred N. *Science and the Modern World*. New York: The Free Press, 1997, p. 39.

[124] Whitehead, op. cit., p. 57.

[125] Whitehead, op. cit., p. 95.

[126] Whitehead, Alfred N. *Science and the Modern World.* New York: The Free Press, 1997, p. 113.

[127] Merculieff, Ilarion (Larry). "The Gifts from the Four Directions," May 20, 2004, URL = <http://www.yesmagazine.org/issues/a-conspiracy-of-hope/the-gifts-from-the-four-directions>.

[128] Campbell, T. Colin and Campbell, Thomas M. II. *The China Study: The Most Comprehensive Study of Nutrition Even Conducted and the Startling Implications for Diet, Weight Loss and Long-term Health.* Dallas: Benbella Books. 2005. This book's main focus is on findings that connect animal protein with disease, based on data that had been previously collected in China and additional research conducted through a twenty-year partnership among Cornell University, Oxford University, and the Chinese Academy of Preventive Medicine. Part IV of that book is dedicated to discussing how the "entire system"—government, science, medicine, industry, media—profits from the control of public information.

[129] Bohm, David. *Causality & Chance in Modern Physics.* University of Pennsylvania, 1957, chapter 2.

[130] Hurtado, Lorenzo Muelas. "Access to the Resources of Biodiversity and Indigenous Peoples," Edmonds Institute occasional paper. Bogota, Colombia, 1998, URL = < www.edmonds-institute.org/muelaseng.html>.

[131] Smolin, Lee. *Three Roads to Quantum Gravity*, New York: Basic Books, 2001, p. 49

[132] Gordon, Oakley. "An Environmental Epistemology of the Andean People of Peru," <http://www.psych.utah.edu/gordon/AndeanResearch/FullPaper.html>.

[133] Peat, F. David. *From Certainty to Uncertainty.* Washington: Joseph Henry Press, 2002.

[134] Aczel, A. *Op. cit.,* p. ix.

[135] Bohm, David. *The Undivided Universe.* New York: Routledge. 1993, p. 2.

[136] Herbert, Nick. "Holistic Physics –or- An Introduction to Quantum Tantra." URL = <http://www.southerncrossreview.org/16/herbert.essay.htm>.

[137] Bertalanffy, Ludwig von. *General System Theory: Foundations, Development, Applications.* New York: George Braziller, 1968, p. ?.

[138] Akwesasne notes, ed. *A Basic Call to Consciousness: The Haudenosaunee Address to the Western World*. Summertown, TN: Native Voices, June 2005, pp. 123-125.

[139] LaViolette, Paul A. *Subquantum Kinetics: A Systems Approach to Physics and Cosmology*. Alexandria, VA: Starlane Publications, 1994, p.1.

[140] Tesla, N. "Man's Greatest Achievement." *New York American*, July 6, 1930.

[141] Quoted by LaViolette, Paul. *Genesis of the Cosmos*. Rochester: Bear & Company, 2004, p. 100

[142] Charles, Prince of Wales. "Sacredness and Sustainability: A Reflection on the 2000 Century," St. James's Palace and the Press Association Ltd., BBC Reith Lectures, May 2000. Prince Charles gave this talk at the conclusion of the annual Reith Lecture series, which took sustainable development as its theme this year. URL = < http://archives.greenmoneyjournal.com/article.mpl?newsletterid=12&articleid=101>.

[143] Merculieff, Ilarion (Larry). "Let Goodness Take Its Place," November 2004, URL = <http://www.healingwiseforum.com/viewtopic.php?f=5&t=9121>.

[144] Statement by Michael Behe in reaction to the judge's decision in the federal court case against the Dover, Pennsylvania School District. See URL = <http://www.discovery.org/scripts/viewDB/filesDB-download.php?command=download&id=697>.

[145] The Iroquois Great Binding Law is posted online at multiple locations. Visit URL = <http://www.fordham.edu/halsall/mod/iroquois.asp>.

[146] This statement is posted online at several locations. A Wikipedia article gives the following reference: Vecsey C. and Venables R. W. (Editors). "An Iroquois Perspective. in American Indian Environments: Ecological Issues in Native American History." New York: Syracuse University Press, pp. 173, 174. URL = <http://en.wikipedia.org/wiki/Seven_generation_sustainability>.

[147] The Seventh Generation Amendment is stated in a brochure available online on February 2, 2013 at URL = <http://www.gp.org/committees/ecoaction/documents/Seventh_Generation_brochure.pdf >.

[148] Smolin. op. cit., pp. 49-50

[149] Churchill, Ward and Morris, Glenn T.. "Key Indian Laws and Cases," in *The State of Native America*, M. Annette Jaimes, ed. Boston: South End Press, 1992, pp. 35-36.

[150] Utter, Jack. "The Discovery Doctrine, the Tribes, and the Truth (Part I)," in *Indian Country Today Media Network*, May 24, 2000, URL = <http://indiancountrytodaymedianetwork.com/ictarchives/2000/05/24/part-i-the-discovery-doctrine-the-tribes-and-the-truth-87045>.

[151] Churchill, Ward and Morris, Glenn T.. "Key Indian Laws and Cases," in *The State of Native America*, M. Annette Jaimes, ed. Boston: South End Press, 1992.

[152] Utter, Jack, *op. cit.*

Bibliography

Aczel, A. 2001. *Entanglement*. New York: Penguin Books.

Aikenhead, G. "Integrating Western and Aboriginal Sciences: Cross-Cultural Science Teaching," in *Research in Science Education*, 2001, vol. 31, no. 3.

Aikenhead, Glen and Michell, Herman. *Bridging Cultures: Indigenous and Scientific Ways of Knowing Nature*. Toronto: Pearson Canada, Inc., 2011.

Aikenhead, Glen. "Rekindling Traditions: Cross-Cultural Science and Technology Project," 2000, URL = <http://www.usask.ca/education/ccstu>.

Akwesasne notes, ed. *A Basic Call to Consciousness: The Haudenosaunee Address to the Western World*. Summertown, TN: Native Voices, June 2005. This publications contains the entire 1977 message (consisting of three position papers) which was published in 1978. The entire 1977 address is available on the Internet at URL= <www.ratical.com/many_worlds/6Nations/BasicCtC.html>.

Angayuqaq Oscar Kawagley and Ray Barnhardt. "Education Indigenous to Place: Western Science Meets Native Reality," retrieved June 12, 2011.

Armstrong, Jeanette. "Sharing One Skin: the Okanagan Community," in *The Case Against Global Economy*, Jerry Mander and Edward Goldsmith (eds.), San Francisco: Sierra Club Books, 1996.

Barbour, Ian G.. *When Science Meets Religion*. New York: HarperCollins, 2000.

Barreiro, José (ed.). *Indian Roots of American Democracy*. Ithaca: Cornell University, 1992.

Berardi, Burns, Duran, Gonzalez-Plaza, Kinley, Robbins, Williams, and Woods. *Handbook for Facilitators: Principles and Adaptation of the Tribal Environmental and Natural Resources Management (TENRM) Model for Tribal Colleges*. Bellingham: Northwest Indian College. 2001. 102 pages.

Berardi, Burns, Duran, Gonzalez-Plaza, Kinley, Robbins, Williams, and Woods. "Profile and rationale for the Tribal Environmental and Natural Resources Management (TENRM) Program at the Northwest Indian College: An innovative, interdisciplinary Native American curriculum." American Indian Culture and Research Journal 26:3 (2002), pp. 47-64.

Bertalanffy, Ludwig von. *General System Theory: Foundations, Development, Applications*. New York: George Braziller, 1968.

Black Elk. *Black Elk Speaks* (as told through John G. Neihardt). Lincoln: University of Nebraska, 1932.

Black Elk. *Black Elk Speaks*. (J. Neihardt, ed.) Lincoln, NE: University of Nebraska Press. 1961.

Bohm, D., Hiley, B, J. 1993. *The Undivided Universe*. London and New York: Routledge.

Bohm, David. *Causality & Chance in Modern Physics*. University of Pennsylvania, 1957.

Bohm, David. *The Special Theory of Relativity*. London and New York: Routledge, 1996.

Bohm, David. *The Undivided Universe*. New York: Routledge. 1993.

Bohm, David. *Wholeness and the Implicate Order*. New York: Routledge. 1980.

Bohm, David and Krishnamurti, Jiddu. *The Ending of Time*.

Bopp, Judie. *The Sacred Tree*. Twin Lakes, WI: Lotus Press, 1984.

Brooks, Drex. *Sweet Medicine: Sites of Indian Massacres, Battlefields, and Treaties*. 1995. Albuquerque: University of New Mexico Press.

Cajete, Gregory A., *Igniting the Sparkle: An Indigenous Science Education Model*. Skyland, NC: Kivaki Press. 1999.

Cajete, Gregory. *Look to the Mountain: An Ecology of Indigenous Education*. Durango, Colorado: Kivaki Press, 1994.

Cajete, Gregory. *Native Science: Natural Laws of Interdependence*. Santa Fe: Clear Light Publishers, 2000.

Callender, Craig and Weingard, Robert. "The Bohmian Model of Quantum Cosmology," *PSA: Proceedings of the Biennial Meeting of the Philosophy of Science Association*, Volume One: Contributed Papers (1994), pp. 218-227, URL = <http://www.jstor.org/stable/193027>.

Campbell, T. Colin and Campbell, Thomas M. II. *The China Study: The Most Comprehensive Study of Nutrition Even Conducted and the Startling Implications for Diet, Weight Loss and Long-term Health*. Dallas: Benbella Books. 2005.

Capra, F. *The Tao of Physics*. Boston: Shambala Publications, Inc., 1985.

Smolin, Lee. *Three Roads to Quantum Gravity*, New York: Basic Books, 2001.

Charles. The Prince of Wales. "Sacredness and Sustainability: A Reflection on the 2000 Century," St. James's Palace and the Press Association Ltd., BBC Reith Lectures, May 2000. Prince Charles gave this talk at the conclusion of the annual Reith Lecture series, which took sustainable development as its theme this year. URL=<http://archives.greenmoneyjournal.com/article.mpl?newsletterid=12&articleid=101>.

Churchill, Ward and Morris, Glenn T.. "Key Indian Laws and Cases," in *The State of Native America*, M. Annette Jaimes, ed. Boston: South End Press, 1992.

Curd, Patricia, "Presocratic Philosophy", *The Stanford Encyclopedia of Philosophy (Spring 2012 Edition)*, Edward N. Zalta (ed.), URL = <http://plato.stanford.edu/archives/spr2012/entries/presocratics/>.

Davis, Wade (2011-05-29). *The Wayfinders: Why Ancient Wisdom Matters in the Modern World* (CBC Massey Lecture). Perseus Books Group. Kindle Edition.

Deloria, Vine, Jr. and Wildcat, Daniel R.. *Power and Place. Power and Place: Indian Education in America*. Golden, Colorado: Fulcrum Resources, 2001.

Deloria, Vine, Jr. *The World We Used to Live In: Remembering the Powers of the Medicine Men*. Golden, Colorado: Fulcrum Publishing, 2006.

Deloria, Vine. *Evolution, Creationism, and Other Modern Myths*. Golden, Colorado: Fulcrum Publishing, 2002.

Duran, Eduardo and Duran, Bonnie. *Native American Post-Colonial Psychology*. State University of New York Press, 1995.

Duran, Phillip H. "On the Cosmic Order of Modern Physics and the Conceptual World of the American Indian," in *World Futures*, Taylor & Francis, 63: 1–27, 2007.

Duran, Phillip H. *Evaluation of Bessel Functions with Imaginary Order for Applications to Certain Boundary-Value Problems*. Final technical report for 1 Jul 69-30 Jun 70, accession number AD0728833, The University of Texas at El Paso, June 1970.

Duran, Phillip H. *Sound Intensity in Atmospheric Shadow Zones Assuming a Constant Velocity Gradient*. Final technical report for 1 Jul-30 Sep 70, accession number AD0728836, The University of Texas at El Paso, December 1970.

Eastman, Charles (Ohiyesa). *The Soul of an Indian*. Kent Nerburn, ed. Novato, California: New World Library, 1993.[1]

Einstein, Albert. *Albert Einstein: Out of My Later Years*. New York: Winds Books. 1956.

Einstein, Albert. *The Meaning of Relativity*. Princeton University Press, 1988. Introduction by Brian Greene, 2005.

Einstein, Albert. *Relativity: The Special and the General Theory*. New York: Crown Publishers, Inc., 1961.

Elgin, Duane. "Scientific Evidence of a Living Conscious Universe," June 8, 2008, URL = <http://www.dreammanifesto.com/scientific-evidence-living-consciousness-universe.html>.

Elgin, Duane. "We Live in a Living Universe," *Working with Oneness*, URL = <http://www.workingwithoneness.org/articles/we-live-living-universe>.

Ereira, Alan. *The Elder Brothers' Warning*. Tairona Heritage Trust, 2009. The book has been re-published by Tairona Heritage Trust (www.taironatrust.org), a British-based NGO that has been working in behalf of the Gonavindua Tairona (the Kogi culture).

Ereira, Alan. *From the Heart of the World: The Elder Brothers' Warning* [VHS], 1998.

---

[1] This edition is based on the original 1911 publication.

Farmelo, Graham. *The Strangest Man: The Hidden Life of Paul Dirac, Mystic of the Atom.* New York: Perseus Books Group, 2009.

Feulner, Edwin, "Roots of Conservative Thought from Burke to Kirk," *First Principles Series* 19 (July 9, 2008), Washington, DC: The Heritage Foundation.

Gordon, Oakley. "An Environmental Epistemology of the Andean People of Peru, 2001," According to Gordon, the translation of the term "paq'o" is difficult to arrive at as there is no exact equivalent in our society. Terms such as healer, spiritual leader, teacher, shaman, mystic seem to apply. Retrieved January 24, 2013. URL = <http://www.psych.utah.edu/gordon/AndeanResearch/FullPaper.html>.

Greene, Brian. *The Fabric of the Cosmos.* New York: Vintage Books, 2005.

Hawken, Paul. *Blessed Unrest: How the Largest Movement in the World Came into Being and Why No One Saw It Coming.* New York: Penguin Group, 2007.

Hawking, S. and Mlodinow, Leonard. *The Grand Design.* Bantam Books, 2010.

Herbert, Nick. "Holistic Physics –or– An Introduction to Quantum Tantra." URL = <http://www.southerncrossreview.org/16/herbert.essay.htm>.

Hurtado, Lorenzo Muelas. "Access to the Resources of Biodiversity and Indigenous Peoples," Edmonds Institute occasional paper. Bogota, Colombia, 1998, URL = < www.edmonds-institute.org/muelaseng.html>

Indian Nations at Risk Task Force. (1991, October). *Indian nations at Risk: An educational strategy for action* (Final report of the

Indian Nations at Risk Task Force). Washington, DC: U.S. Department of Education.

Friends Committee on National Legislation. *Indian Report*, Washington DC, issue #92, Fall 2012.

"Inuit Observations on Climate Change," Community Adaptation and Sustainable Livelihoods, URL = <www.iisd.org/casl/projects/inuitobs.htm>.

Isaacson, Walter. *Einstein: His Life and Universe*. New York: Simon and Schuster, 2007.

Johnson, P. 2000. *The Wedge of Truth*. Downers Grove, Illinois: InterVarsity Press, 2000.

Josephson, B., Rubik, B. "The Challenge of Consciousness Research," URL = <http://www.tcm.phy.cam.ac.uk/~bdj10/mm/articles/athens.txt>.

Josephson, Brian D. and Utts, Jessica. "The Paranormal: The Evidence and its Implications for Consciousness." URL = <http://www.tcm.phy.cam.ac.uk/~bdj10/psi/tucson.html>.

Kafatos, M. and Nadeau, R. *The Conscious Universe: Part and Whole in Modern Physical Theory*. New York: Springer-Verlag, 1990.

Kafatos, M. and Nadeau, R. *The Conscious Universe: Parts and Wholes in Modern Physical Reality*. New York: Springer, 2000.

Kakalios, James. *The Amazing Story of Quantum Mechanics*. New York: Gotham Books, 2010.

Kawagley, Angayuqaq O. and Barnhardt, Ray. "Education Indigenous to Place: Western Science Meets Native Reality," URL = <www.ankn.uaf.edu/Curriculum/Articles/BarnhardtKawagley/EIP.html>.

Keoke, Emory Dean, and Porterfield, Kay Marie. *Encyclopedia of American Indian Contributions to the World: 15,000 Years of Inventions and Innovations,* Facts on File Library of American History, 2002.[2]

Kidwell, Clara Sue, Noley, Homer, and Tinker, George E. "Tink". *A Native American Theology.* Maryknoll, New York: Orbis Books, 2001.

Kingsley, Peter. *Ancient Philosophy, Mystery, and Magic: Empedocles and Pythagorean Tradition.* Oxford: Oxford University Press, 1995.

Kingsley, Peter. *A Story Waiting to Pierce You: Mongolia, Tibet and the Destiny of the Western World.* Point Reyes, California: The Golden Sufi Center, 2010.

Kingsley, Peter. "Original Instructions: An Interview with Peter Kingsley," in *Quest,* Volume 99, Number 3, Summer 2011.

LaViolette, Paul A. *Genesis of the Cosmos.* Rochester, Vermont: Bear & Company, 2004.

LaViolette, Paul A. *Subquantum Kinetics: A Systems Approach to Physics and Cosmology.* Alexandria, VA: Starlane Publications, 1994.

Lemay, Connie. "A Brief History of American Indian Military Service," in *Indian Country Today Media Network,* May 28, 2012, URL = <http://indiancountrytodaymedianetwork.com/article/a-brief-history-of-american-indian-military-service-115318>.

Lindberg, David C. *The Beginnings of Western Science.* University of Chicago Press, 2007.

---

[2] A shorter edition for younger readers, by the same title, was published in 2005.

Lovelock. *Gaia: A New Look at Life on Earth*. Oxford: University Press, 2000.

Lyons, Oren. *Oren Lyons: The Faithkeeper*. The Moyers Collection [VHS] (1991).

Lyons, Oren. "Our Mother Earth," *Parabola*, Volume 7, No. 1, Winter, 1984.

Men, Hunbatz. *Secrets of Mayan Science/Religion*. Santa Fe: Bear & Company. 1990.

Merculieff, Ilarion (Larry). "The Gifts from the Four Directions," May 20, 2004, URL = <http://www.yesmagazine.org/issues/a-conspiracy-of-hope/the-gifts-from-the-four-directions>.

Merculieff, Ilarion (Larry). "Let Goodness Take Its Place," November 2004, URL = <http://www.healingwiseforum.com/viewtopic.php?f=5&t=9 121>

Mih, Walter. *La Fascinante Vida y Teoría de Albert Einstein*, (*The Fascinating Life and Theory of Albert Einstein*). Francisco Manzo-Robledo (translator), Mexico: LIBROS PARA TODOS, S.A.

Mohawk, John, ed. *A Basic Call to Consciousness: The Haudenosaunee Address to the Western World*. Rooseveltown: Akwesasne Notes, 1978

Momaday, N. Scott. "An American Land Ethic," in *The Man Made of Words: Essays, Stories, and Passages*. New York: St. Martin's Press, 1998.

Mooney, Chris. *The Republican Brain: The Science of Why They Deny Science and Reality*. Wiley, 2012.

Napoleon, Harold. *Yuuyaraq: The Way of the Human Being*. Fairbanks: Alaska Native Knowledge Network, 1996.

Needleman, Jacob. *The American Soul: Rediscovering the Wisdom of the Founders*. New York: Tarcher, 2003.

Nelson, Melissa K., ed. *Original Instructions*. Rochester: Bear & Company, 2008.

Nichols, Edwin J. "The Philosophical Aspects of Cultural Differences." Cultural Bias Methods of Analysis Conference speech on DVD, Men and Women of Respect Historical Society, URL = www.MenofRespect.com.

Peat, F. David. *From Certainty to Uncertainty*. Washington: Joseph Henry Press, 2002.

Petersen, A. "The Philosophy of Niels Bohr," *Bulletin of the Atomic Scientists*. September 1963.

Phillips, John L. "A Tribal College Land Grant Perspective: Changing the Conversation." Journal of American Indian Education. Volume 42, Issue 1. 2003.

Pierotti, R. and Wildcat, D. "Traditional Knowledge, Ecology, and Evolution," pre-publication manuscript.

Rael, Joseph. *Ceremonies of the Living Spirit*. Tulsa/San Francisco: Council Oaks Books, 1998.

Rael, Joseph. *In the House of Shattering Light*. Special Edition. 2002.

Ramos, Jorge. *A Country for All: An Immigrant Manifesto*. New York: Vintage Books, 2010.

Sando, Joe S. *Nee Hemish: A History of Jemez Pueblo*. Santa Fe: Clear Light Publishing.

Schaefer, Carol. *Grandmothers Counsel the World: Women Elders Offer Their Vision for Our Planet*. Foreword by Winona LaDuke. Boston: Trumpeter Books, 2006. Kindle Edition.

Their website is at URL = <
http://www.grandmotherscouncil.org>.

Schultz, Paul and Tinker, Tink. "Rivers of Life" in *Native and Christian: Indigenous Voices on Religious Identity in the United States and Canada*. James Treat, ed. New York: Routledge, 1996.

Smolin, Lee. *The Trouble with Physics*. Boston: Houghton Mifflin Company, 2006.

Smolin, Lee. *Three Roads to Quantum Gravity*, New York: Basic Books, 2001.

Stannard, David E. *American Holocaust: Columbus and the Conquest of the New World*. New York: Oxford University Press, 1992.

Tesla, Nikola. "Man's Greatest Achievement" in *Universal Laws Never Before Revealed: Keely's Secrets: Understanding and Using the Science of Sympathetic Vibration*. Santa Fe: The Message Company, Revised Edition, April 1, 2007..

The Mystery of Chaco Canyon. Produced by Anna Sofaer (The Solstice Project, 1999, VHS), Bullfrog Films.

Thornton, Russell. *American Indian Holocaust and Survival*. Norman: The University of Oklahoma Press, 1990.

Utter, Jack. *American Indians: Answers to Today's Questions*. Norman: The University of Oklahoma Press, 2002.

Utter, Jack. "The Discovery Doctrine, the Tribes, and the Truth (Part I)," in *Indian Country Today Media Network*, May 24, 2000, URL = <
http://indiancountrytodaymedianetwork.com/ictarchives/2000/05/24/part-i-the-discovery-doctrine-the-tribes-and-the-truth-87045>.

Utter, Jack. "The Discovery Doctrine, the tribes, and the truth (Part II)," in *Indian Country Today*. June 7, 2000.

Vecsey C. and Venables R. W. (editors). "An Iroquois Perspective," in *American Indian Environments: Ecological Issues in Native American History*. New York: Syracuse University Press, 1980.

Whitehead, Alfred N. *Science and the Modern World*. New York: The Free Press, 1997.

Wolfe, F. A. *Taking the Quantum Leap*. New York: Harper & Row, 1989.

Index

Four Directions · 330
fundamental force · 158, 304
   electromagnetic · 63, 126
   force carrier · 126
   gravitational · 69, 117, 121, 126, 151, 155
   strong nuclear · 126
   weak nuclear · 126

G

Galilei, Galileo · 147, 163, 242, 243, 244, 245, 261, 264
General Systems theory · 289, 290
genes · 167, 265
Genesis · 84, 238, 305, 324, 331
genetic · 45, 161
geology · 199, 235
geometry · 60, 62, 144, 150, 153, 154, 155, 156, 241, 269, 272
   Euclidean · 144, 153, 241
   non-Euclidean · 144
   Riemannian · 61, 154, 155, 156
global · xii, 7, 9, 10, 16, 17, 25, 38, 56, 65, 183, 187
Gordon, Oakley · 275, 330
Göttingen University · 11
graduate · xix, xxiii, 92, 204, 220, 243
Greece · xxiv, 2, 13, 30, 46, 58, 99, 130, 171, 173, 183, 225, 240, 245
Greeks · xxiv, 2, 13, 15, 36, 46, 58, 59, 62, 71, 73, 87, 162, 163, 173, 217, 225, 227, 228, 239, 242, 248, 250, 299, 314
   Presocratics · 36, 58, 60, 226, 227, 228, 229, 230
GST · 289, 290

H

harmonious · 22, 223, 306, 319
harmony · xii, 11, 27, 81, 100, 137, 182, 192, 236, 299
Haudenosaunee (Iroquois) · 44, 45, 83, 185, 194, 196, 323, 327, 331
heart · xi, xvi, xxiii, 1, 19, 28, 43, 72, 163, 299, 301, 308

Heisenberg, Werner · xiv, 65, 111, 112, 114, 115, 116, 117, 118, 166, 273, 276, 278, 279, 280
high school · xix, 41, 53, 212
holism · xii, xvi, xxv, 2, 14, 15, 17, 23, 37, 59, 131, 133, 134, 163, 174, 212, 213, 217, 218, 219, 220, 226, 228, 239, 249, 251, 276, 285, 288, 289, 292, 293, 299, 310, 318, 330
Hollywood · 36, 74
Hopi · 18, 44, 303
Huichol · 44
human (humanity) · xii, xiii, 1, 4, 8, 10, 13, 15, 18, 21, 22, 24, 25, 28, 29, 30, 35, 37, 38, 43, 53, 54, 62, 63, 67, 68, 71, 72, 73, 77, 83, 84, 98, 99, 110, 114, 120, 127, 129, 131, 140, 145, 166, 173, 174, 180, 181, 182, 187, 188, 190, 191, 192, 194, 195, 197, 198, 199, 204, 209, 211, 212, 213, 216, 218, 219, 220, 221, 223, 224, 237, 240, 245, 247, 248, 251, 257, 259, 262, 264, 268, 276, 284, 288, 289, 292, 294, 295, 299, 301, 303, 309, 316, 319, 327, 329
humility · 11, 81, 84, 131, 234, 240
hypothesis · 54, 91, 93, 246, 259, 282, 296, 306

I

ideal · 35, 135, 223
identity · xxi, xxiii, 32, 52, 75, 77, 78, 211, 263, 302, 318
ideology · 15, 181, 311
indigenization · xxii, 4
Industrial Revolution · 6, 163, 247
intellect · xxiii, 18, 88, 165, 171, 177, 178, 199, 209, 221, 224, 227, 229, 237, 239, 243, 248, 251, 256, 264, 278, 293, 317, 319
International · xvii, 35, 42, 44, 77, 78
intuition · 19, 95, 102, 130, 199, 201, 203, 233, 249
intuitive · 102, 123
Inuit · 43, 194, 328

urban · 40

V

validation · xxii, xxiv, 20, 49, 52,
   56, 59, 73, 82, 98, 102, 134, 143,
   144, 153, 165, 169, 191, 193,
   194, 201, 203, 212, 214, 218,
   252, 259, 260, 267, 272, 278,
   294, 304, 312
Van Allen, James · xix
Vesalius · 242
visual · 247

W

water · 7, 46, 88, 100, 101, 105,
   109, 150, 181, 190, 199, 200,
   230, 232, 256, 279, 293, 301,
   302, 307, 308
wave-particle duality · 103, 115,
   205, 266, 280
web · 3, 4, 15, 22, 38, 45, 56, 97,
   128, 142, 195, 206, 207, 269,
   272, 274, 275, 277, 285, 287,
   299, 309, 310
web of life · 4, 15, 22, 38, 45, 128,
   195, 277, 299, 309, 310
Whitehead, Alfred North · 30, 244,
   245, 246, 249, 296, 322, 329,
   330

whole · xxii, 3, 10, 12, 15, 21, 23,
   50, 54, 55, 84, 96, 97, 98, 104,
   111, 123, 129, 137, 159, 163,
   165, 169, 173, 174, 184, 196,
   199, 201, 205, 206, 207, 209,
   212, 214, 217, 218, 219, 223,
   234, 246, 252, 257, 259, 265,
   267, 272, 273, 276, 283, 285,
   289, 292, 296, 299, 307, 316,
   318
wholeness · xii, xv, 12, 22, 50, 56,
   64, 123, 131, 201, 205, 206, 207,
   212, 214, 251, 252, 262, 267,
   273, 274, 275, 277, 282, 285,
   289, 290, 301, 309
wind · xx, 6, 7, 190, 302
wisdom · xxv, xxvi, 2, 5, 16, 20, 25,
   28, 29, 31, 89, 181, 182, 183,
   185, 196, 197, 198, 202, 212,
   221, 228, 299, 300, 318
worldview · xii, xiii, xv, xxv, 4, 9,
   16, 19, 23, 25, 28, 42, 47, 52, 53,
   59, 71, 73, 114, 123, 129, 133,
   168, 169, 173, 182, 193, 205,
   212, 216, 217, 219, 224, 240,
   246, 248, 249, 257, 268, 276,
   314

Z

zoology · 228, 232, 244

www.ingramcontent.com/pod-product-compliance
Lightning Source LLC
Chambersburg PA
CBHW050449270326
41927CB00009B/1673